RILEM State-of-the-Art Reports

RILEM STATE-OF-THE-ART REPORTS
Volume 50

RILEM, The International Union of Laboratories and Experts in Construction Materials, Systems and Structures, founded in 1947, is a non-governmental scientific association whose goal is to contribute to progress in the construction sciences, techniques and industries, essentially by means of the communication it fosters between research and practice. RILEM's focus is on construction materials and their use in building and civil engineering structures, covering all phases of the building process from manufacture to use and recycling of materials. More information on RILEM and its previous publications can be found on www.RILEM.net.

The RILEM State-of-the-Art Reports (STAR) are produced by the Technical Committees. They represent one of the most important outputs that RILEM generates – high level scientific and engineering reports that provide cutting edge knowledge in a given field. The work of the TCs is one of RILEM's key functions.

Members of a TC are experts in their field and give their time freely to share their expertise. As a result, the broader scientific community benefits greatly from RILEM's activities.

RILEM's stated objective is to disseminate this information as widely as possible to the scientific community. RILEM therefore considers the STAR reports of its TCs as of highest importance, and encourages their publication whenever possible.

The information in this and similar reports is mostly pre-normative in the sense that it provides the underlying scientific fundamentals on which standards and codes of practice are based. Without such a solid scientific basis, construction practice will be less than efficient or economical.

It is RILEM's hope that this information will be of wide use to the scientific community.

Indexed in SCOPUS, Google Scholar and SpringerLink.

Geert De Schutter · Jay Sanjayan
Editors

Active Control of Properties of Fresh and Hardening Cementitious Materials

State-of-the-Art Report of the RILEM Technical Committee 317-ACP

Springer

Editors
Geert De Schutter
Department of Structural Engineering
and Building Materials, Faculty
of Engineering and Architecture
Ghent University
Ghent, Belgium

Jay Sanjayan
School of Engineering
Swinburne University of Technology
Melbourne, VIC, Australia

ISSN 2213-204X ISSN 2213-2031 (electronic)
RILEM State-of-the-Art Reports
ISBN 978-3-032-30339-4 ISBN 978-3-032-30340-0 (eBook)
https://doi.org/10.1007/978-3-032-30340-0

This Springer imprint is published by the registered company Springer Nature Switzerland AG
The registered company address is: Gewerbestrasse 11, 6330 Cham, Switzerland

Contributors—RILEM Technical Committee 317-ACP

Chairman

Geert De Schutter, Belgium

Deputy Chairman

Jay Sanjayan, Australia

Active Members

Akilesh Ramesh, Australia,
Angela Tetteh Tawiah, Italy,
Angel de la Rosa Velasco, Spain
Anil Kumar Kannauzia, India
Dali Bonar, UK
Dengwu Jiao, China
Didier Snoeck, Belgium
Dimitri Feys, USA
Dimitrios G. Aggelis, Belgium
Eleni Korda, Belgium
Gashaw Abebaw Adanu, Ethiopia
Imene Abidi, Italy
Kolawole A. Olonade, Nigeria
Luise Göbel, Germany
Manu Kurungod Mohan, Belgium
Mert Yucel Yardimci, Türkiye

Noor Yaseen, Pakistan
Ojas Arun Chaudhari, Sweden
Onur Ozturk, USA
Qiang Yuan, China
Paula Heik, Germany
Ponsian Mwombeki Robert, Cyprus
Shravan Muthukrishnan, Germany
Sriramya Nair, USA
Swapnil Balasaheb Ghodke, India
Thomas Kränkel, Germany
Williap Peter Boshoff, South Africa
Yaxin Tao, China
Ye Qian, China
Yi Zhang, Belgium
Yiyuan Zhang, Belgium

Preface

Concrete practice is doing a marvellous job, day by day, worldwide, resulting in common to record-breaking concrete structures. Building with concrete seems obvious, straightforward, easy. Nevertheless, processing concrete can be very challenging. Variations in properties and dosage of constituents as well as variations in environmental and practical conditions, can cause the concrete performance to deviate from the intended performance. In traditional approaches, the workers and site engineers do not have many options to adjust the concrete while it is being processed.

As recently fundamentally introduced and studied, the concept of Active Control potentially offers advanced options to intervene in order to adjust the concrete performance to match the desired performance. As defined in Chapter 'Introduction to Active Control of Properties of Fresh and Hardening Cementitious Materials', Active Control of properties of cementitious materials refers to the action of controlling some specific properties (e.g. rheology and stiffening) by applying an external trigger signal that induces some response in the cementitious material, with a modification of properties or behaviour as a result. In the ideal active control process, the intervention is based on observations and/or measurements (quantified), giving the necessary information to take a decision regarding required intervention. A feed-back loop between measurement and signal seems an optimal situation.

This document offers the first State-of-the-Art Report on Active Control of Properties of Cementitious Materials, as prepared within RILEM Technical Committee 317-ACP. The activities of the TC started in September 2023, with the inaugural meeting held during the RILEM Annual Week in Vancouver. During about 2 years, the full State-of-the-Art Report was discussed, outlined, drafted, internally reviewed, and put into final format. The cut-off date for inclusion of published papers and reports was set at 1 January 2025. Some papers formally published after this date have been included as well, but the main content of the report is based on Active Control literature published before this cut-off date.

It is clear that this subfield of Active Control has gained significant momentum in the last year(s). Many groups worldwide are currently working on scientific projects studying different concepts and different applications of Active Control. This is great to see! All of this will contribute to a further advancement of our industry. The downside is that this State-of-the-Art Report will soon be outdated, with new information and new insights going way beyond the current content. Consider this Report as a first comprehensive document, defining the concepts of Active Control and introducing potential applications. Within several years, we might all be positively surprised by the progress that will have been made by then. We will then, hopefully, look back to this Report as a pioneering contribution to a new and fast developing subdomain. We currently learn how to walk. Within several years, we can hopefully run a marathon.

Upon the publication of the State-of-the-Art Report, I am very grateful to all active TC members (see higher) for their very valuable and tremendous committee work! I also acknowledge the observing members for their interest: A. V. Rahul, India; Alexander Mezhov, Germany; Ana Brunčič, Slovenia; Anabela Guadalupe Guilarducci, Argentina; Ashirbad Satapathy, India; Bernd Rottmann, Germany; Caleb Olutunde, Nigeria; Chris Wilkes, UK; Huu May Nguyen, Japan; Jennifer Anette Canul Polanco, Denmark; Jing Yu, China; Khadija El Cheikh, Belgium; Kopitha Kirushnapillai, Australia; Marimuthu Krishnan, India; Muralidhar Kamath, India; Musbau Busari, Nigeria; Paulo Matos, Brazil; Raissa Douglas Ferron, USA; Robin De Schryver, Belgium; Shiwei Yu, Belgium; Siamak Fakhrayee Nejad, Iran; Sofiane Amroun, USA; Sribhanupratap Rathod Ramawat, Switzerland; Tafadzwa Ronald Muzenda, Germany; Teresa Liberto, Germany; Vinh Dao, Australia; Wasiu Ajagbe, Nigeria; Yanwei Wang, Kazakhstan; Yu Chen, China.

It was my honour and pleasure to chair this Technical Committee, together with Jay Sanjayan, whom I am very thankful for taking the role of deputy-chair. I hope this Report is of value to you, whether you are a student, a scientist or a practitioner. I look forward to new studies, inventions and developments in this subdomain of Active Control of Cementitious Materials. Any further active contribution to increase reliability, safety and sustainability in our field will be largely appreciated.

Ghent, Belgium
Melbourne, Australia
December 2025

Geert De Schutter
Jay Sanjayan

RILEM Publications

The following list presents the global offer of RILEM Publications, sorted by series. Each publication is available in printed version and/or in online version.

RILEM Proceedings (PRO)

PRO 1: Durability of High Performance Concrete (ISBN: 2-912143-03-9; e-ISBN: 2-351580-12-5; e-ISBN: 2351580125); *Ed. H. Sommer*
PRO 2: Chloride Penetration into Concrete (ISBN: 2-912143-00-04; e-ISBN: 2912143454); *Eds. L.-O. Nilsson and J.-P. Ollivier*
PRO 3: Evaluation and Strengthening of Existing Masonry Structures (ISBN: 2-912143-02-0; e-ISBN: 2351580141); *Eds. L. Binda and C. Modena*
PRO 4: Concrete: From Material to Structure (ISBN: 2-912143-04-7; e-ISBN: 2351580206); *Eds. J.-P. Bournazel and Y. Malier*
PRO 5: The Role of Admixtures in High Performance Concrete (ISBN: 2-912143-05-5; e-ISBN: 2351580214); *Eds. J. G. Cabrera and R. Rivera-Villarreal*
PRO 6: High Performance Fiber Reinforced Cement Composites—HPFRCC 3 (ISBN: 2-912143-06-3; e-ISBN: 2351580222); *Eds. H. W. Reinhardt and A. E. Naaman*
PRO 7: 1st International RILEM Symposium on Self-Compacting Concrete (ISBN: 2-912143-09-8; e-ISBN: 2912143721); *Eds. Å. Skarendahl and Ö. Petersson*
PRO 8: International RILEM Symposium on Timber Engineering (ISBN: 2-912143-10-1; e-ISBN: 2351580230); *Ed. L. Boström*
PRO 9: 2nd International RILEM Symposium on Adhesion between Polymers and Concrete ISAP '99 (ISBN: 2-912143-11-X; e-ISBN: 2351580249); *Eds. Y. Ohama and M. Puterman*
PRO 10: 3rd International RILEM Symposium on Durability of Building and Construction Sealants (ISBN: 2-912143-13-6; e-ISBN: 2351580257); *Ed. A. T. Wolf*

PRO 11: 4th International RILEM Conference on Reflective Cracking in Pavements (ISBN: 2-912143-14-4; e-ISBN: 2351580265); *Eds. A. O. Abd El Halim, D. A. Taylor and El H. H. Mohamed*
PRO 12: International RILEM Workshop on Historic Mortars: Characteristics and Tests (ISBN: 2-912143-15-2; e-ISBN: 2351580273); *Eds. P. Bartos, C. Groot and J. J. Hughes*
PRO 13: 2nd International RILEM Symposium on Hydration and Setting (ISBN: 2-912143-16-0; e-ISBN: 2351580281); *Ed. A. Nonat*
PRO 14: Integrated Life-Cycle Design of Materials and Structures—ILCDES 2000 (ISBN: 951-758-408-3; e-ISBN: 235158029X); (ISSN: 0356-9403); *Ed. S. Sarja*
PRO 15: Fifth RILEM Symposium on Fibre-Reinforced Concretes (FRC)—BEFIB'2000 (ISBN: 2-912143-18-7; e-ISBN: 291214373X); *Eds. P. Rossi and G. Chanvillard*
PRO 16: Life Prediction and Management of Concrete Structures (ISBN: 2-912143-19-5; e-ISBN: 2351580303); *Ed. D. Naus*
PRO 17: Shrinkage of Concrete—Shrinkage 2000 (ISBN: 2-912143-20-9; e-ISBN: 2351580311); *Eds. V. Baroghel-Bouny and P.-C. Aïtcin*
PRO 18: Measurement and Interpretation of the On-Site Corrosion Rate (ISBN: 2-912143-21-7; e-ISBN: 235158032X); *Eds. C. Andrade, C. Alonso, J. Fullea, J. Polimon and J. Rodriguez*
PRO 19: Testing and Modelling the Chloride Ingress into Concrete (ISBN: 2-912143-22-5; e-ISBN: 2351580338); *Eds. C. Andrade and J. Kropp*
PRO 20: 1st International RILEM Workshop on Microbial Impacts on Building Materials (CD 02) (e-ISBN 978-2-35158-013-4); *Ed. M. Ribas Silva*
PRO 21: International RILEM Symposium on Connections between Steel and Concrete (ISBN: 2-912143-25-X; e-ISBN: 2351580346); *Ed. R. Eligehausen*
PRO 22: International RILEM Symposium on Joints in Timber Structures (ISBN: 2-912143-28-4; e-ISBN: 2351580354); *Eds. S. Aicher and H.-W. Reinhardt*
PRO 23: International RILEM Conference on Early Age Cracking in Cementitious Systems (ISBN: 2-912143-29-2; e-ISBN: 2351580362); *Eds. K. Kovler and A. Bentur*
PRO 24: 2nd International RILEM Workshop on Frost Resistance of Concrete (ISBN: 2-912143-30-6; e-ISBN: 2351580370); *Eds. M. J. Setzer, R. Auberg and H.-J. Keck*
PRO 25: International RILEM Workshop on Frost Damage in Concrete (ISBN: 2-912143-31-4; e-ISBN: 2351580389); *Eds. D. J. Janssen, M. J. Setzer and M. B. Snyder*
PRO 26: International RILEM Workshop on On-Site Control and Evaluation of Masonry Structures (ISBN: 2-912143-34-9; e-ISBN: 2351580141); *Eds. L. Binda and R. C. de Vekey*
PRO 27: International RILEM Symposium on Building Joint Sealants (CD03; e-ISBN: 235158015X); *Ed. A. T. Wolf*
PRO 28: 6th International RILEM Symposium on Performance Testing and Evaluation of Bituminous Materials—PTEBM'03 (ISBN: 2-912143-35-7; e-ISBN: 978-2-912143-77-8); *Ed. M. N. Partl*

PRO 29: 2nd International RILEM Workshop on Life Prediction and Ageing Management of Concrete Structures (ISBN: 2-912143-36-5; e-ISBN: 2912143780); *Ed. D. J. Naus*

PRO 30: 4th International RILEM Workshop on High Performance Fiber Reinforced Cement Composites—HPFRCC 4 (ISBN: 2-912143-37-3; e-ISBN: 2912143799); *Eds. A. E. Naaman and H. W. Reinhardt*

PRO 31: International RILEM Workshop on Test and Design Methods for Steel Fibre Reinforced Concrete: Background and Experiences (ISBN: 2-912143-38-1; e-ISBN: 2351580168); *Eds. B. Schnütgen and L. Vandewalle*

PRO 32: International Conference on Advances in Concrete and Structures 2 vol. (ISBN (set): 2-912143-41-1; e-ISBN: 2351580176); *Eds. Ying-shu Yuan, Surendra P. Shah and Heng-lin Lü*

PRO 33: 3rd International Symposium on Self-Compacting Concrete (ISBN: 2-912143-42-X; e-ISBN: 2912143713); *Eds. Ó. Wallevik and I. Níelsson*

PRO 34: International RILEM Conference on Microbial Impact on Building Materials (ISBN: 2-912143-43-8; e-ISBN: 2351580184); *Ed. M. Ribas Silva*

PRO 35: International RILEM TC 186-ISA on Internal Sulfate Attack and Delayed Ettringite Formation (ISBN: 2-912143-44-6; e-ISBN: 2912143802); *Eds. K. Scrivener and J. Skalny*

PRO 36: International RILEM Symposium on Concrete Science and Engineering—A Tribute to Arnon Bentur (ISBN: 2-912143-46-2; e-ISBN: 2912143586); *Eds. K. Kovler, J. Marchand, S. Mindess and J. Weiss*

PRO 37: 5th International RILEM Conference on Cracking in Pavements—Mitigation, Risk Assessment and Prevention (ISBN: 2-912143-47-0; e-ISBN: 2912143764); *Eds. C. Petit, I. Al-Qadi and A. Millien*

PRO 38: 3rd International RILEM Workshop on Testing and Modelling the Chloride Ingress into Concrete (ISBN: 2-912143-48-9; e-ISBN: 2912143578); *Eds. C. Andrade and J. Kropp*

PRO 39: 6th International RILEM Symposium on Fibre-Reinforced Concretes—BEFIB 2004 (ISBN: 2-912143-51-9; e-ISBN: 2912143748); *Eds. M. Di Prisco, R. Felicetti and G. A. Plizzari*

PRO 40: International RILEM Conference on the Use of Recycled Materials in Buildings and Structures (ISBN: 2-912143-52-7; e-ISBN: 2912143756); *Eds. E. Vázquez, Ch. F. Hendriks and G. M. T. Janssen*

PRO 41: RILEM International Symposium on Environment-Conscious Materials and Systems for Sustainable Development (ISBN: 2-912143-55-1; e-ISBN: 2912143640); *Eds. N. Kashino and Y. Ohama*

PRO 42: SCC'2005—China: 1st International Symposium on Design, Performance and Use of Self-Consolidating Concrete (ISBN: 2-912143-61-6; e-ISBN: 2912143624); *Eds. Zhiwu Yu, Caijun Shi, Kamal Henri Khayat and Youjun Xie*

PRO 43: International RILEM Workshop on Bonded Concrete Overlays (e-ISBN: 2-912143-83-7); *Eds. J. L. Granju and J. Silfwerbrand*

PRO 44: 2nd International RILEM Workshop on Microbial Impacts on Building Materials (CD11) (e-ISBN: 2-912143-84-5); *Ed. M. Ribas Silva*

PRO 45: 2nd International Symposium on Nanotechnology in Construction, Bilbao (ISBN: 2-912143-87-X; e-ISBN: 2912143888); *Eds. Peter J. M. Bartos, Yolanda de Miguel and Antonio Porro*

PRO 46: Concrete Life'06—International RILEM-JCI Seminar on Concrete Durability and Service Life Planning: Curing, Crack Control, Performance in Harsh Environments (ISBN: 2-912143-89-6; e-ISBN: 291214390X); *Ed. K. Kovler*

PRO 47: International RILEM Workshop on Performance Based Evaluation and Indicators for Concrete Durability (ISBN: 978-2-912143-95-2; e-ISBN: 9782912143969); *Eds. V. Baroghel-Bouny, C. Andrade, R. Torrent and K. Scrivener*

PRO 48: 1st International RILEM Symposium on Advances in Concrete through Science and Engineering (e-ISBN: 2-912143-92-6); *Eds. J. Weiss, K. Kovler, J. Marchand, and S. Mindess*

PRO 49: International RILEM Workshop on High Performance Fiber Reinforced Cementitious Composites in Structural Applications (ISBN: 2-912143-93-4; e-ISBN: 2912143942); *Eds. G. Fischer and V. C. Li*

PRO 50: 1st International RILEM Symposium on Textile Reinforced Concrete (ISBN: 2-912143-97-7; e-ISBN: 2351580087); *Eds. Josef Hegger, Wolfgang Brameshuber and Norbert Will*

PRO 51: 2nd International Symposium on Advances in Concrete through Science and Engineering (ISBN: 2-35158-003-6; e-ISBN: 2-35158-002-8); *Eds. J. Marchand, B. Bissonnette, R. Gagné, M. Jolin and F. Paradis*

PRO 52: Volume Changes of Hardening Concrete: Testing and Mitigation (ISBN: 2-35158-004-4; e-ISBN: 2-35158-005-2); *Eds. O. M. Jensen, P. Lura and K. Kovler*

PRO 53: High Performance Fiber Reinforced Cement Composites—HPFRCC5 (ISBN: 978-2-35158-046-2; e-ISBN: 978-2-35158-089-9); *Eds. H. W. Reinhardt and A. E. Naaman*

PRO 54: 5th International RILEM Symposium on Self-Compacting Concrete (ISBN: 978-2-35158-047-9; e-ISBN: 978-2-35158-088-2); *Eds. G. De Schutter and V. Boel*

PRO 55: International RILEM Symposium Photocatalysis, Environment and Construction Materials (ISBN: 978-2-35158-056-1; e-ISBN: 978-2-35158-057-8); *Eds. P. Baglioni and L. Cassar*

PRO 56: International RILEM Workshop on Integral Service Life Modelling of Concrete Structures (ISBN 978-2-35158-058-5; e-ISBN: 978-2-35158-090-5); *Eds. R. M. Ferreira, J. Gulikers and C. Andrade*

PRO 57: RILEM Workshop on Performance of cement-based materials in aggressive aqueous environments (e-ISBN: 978-2-35158-059-2); *Ed. N. De Belie*

PRO 58: International RILEM Symposium on Concrete Modelling—CONMOD'08 (ISBN: 978-2-35158-060-8; e-ISBN: 978-2-35158-076-9); *Eds. E. Schlangen and G. De Schutter*

PRO 59: International RILEM Conference on On Site Assessment of Concrete, Masonry and Timber Structures—SACoMaTiS 2008 (ISBN set: 978-2-35158-061-5; e-ISBN: 978-2-35158-075-2); *Eds. L. Binda, M. di Prisco and R. Felicetti*

PRO 60: Seventh RILEM International Symposium on Fibre Reinforced Concrete: Design and Applications—BEFIB 2008 (ISBN: 978-2-35158-064-6; e-ISBN: 978-2-35158-086-8); *Ed. R. Gettu*

PRO 61: 1st International Conference on Microstructure Related Durability of Cementitious Composites 2 vol., (ISBN: 978-2-35158-065-3; e-ISBN: 978-2-35158-084-4); *Eds. W. Sun, K. van Breugel, C. Miao, G. Ye and H. Chen*
PRO 62: NSF/ RILEM Workshop: In-situ Evaluation of Historic Wood and Masonry Structures (e-ISBN: 978-2-35158-068-4); *Eds. B. Kasal, R. Anthony and M. Drdácký*
PRO 63: Concrete in Aggressive Aqueous Environments: Performance, Testing and Modelling, 2 vol., (ISBN: 978-2-35158-071-4; e-ISBN: 978-2-35158-082-0); *Eds. M. G. Alexander and A. Bertron*
PRO 64: Long Term Performance of Cementitious Barriers and Reinforced Concrete in Nuclear Power Plants and Waste Management—NUCPERF 2009 (ISBN: 978-2-35158-072-1; e-ISBN: 978-2-35158-087-5); *Eds. V. L'Hostis, R. Gens and C. Gallé*
PRO 65: Design Performance and Use of Self-consolidating Concrete—SCC'2009 (ISBN: 978-2-35158-073-8; e-ISBN: 978-2-35158-093-6); *Eds. C. Shi, Z. Yu, K. H. Khayat and P. Yan*
PRO 66: 2nd International RILEM Workshop on Concrete Durability and Service Life Planning—ConcreteLife'09 (ISBN: 978-2-35158-074-5; ISBN: 978-2-35158-074-5); *Ed. K. Kovler*
PRO 67: Repairs Mortars for Historic Masonry (e-ISBN: 978-2-35158-083-7); *Ed. C. Groot*
PRO 68: Proceedings of the 3rd International RILEM Symposium on 'Rheology of Cement Suspensions such as Fresh Concrete (ISBN 978-2-35158-091-2; e-ISBN: 978-2-35158-092-9); *Eds. O. H. Wallevik, S. Kubens and S. Oesterheld*
PRO 69: 3rd International PhD Student Workshop on 'Modelling the Durability of Reinforced Concrete (ISBN: 978-2-35158-095-0); *Eds. R. M. Ferreira, J. Gulikers and C. Andrade*
PRO 70: 2nd International Conference on 'Service Life Design for Infrastructure' (ISBN set: 978-2-35158-096-7, e-ISBN: 978-2-35158-097-4); *Eds. K. van Breugel, G. Ye and Y. Yuan*
PRO 71: Advances in Civil Engineering Materials—The 50-year Teaching Anniversary of Prof. Sun Wei' (ISBN: 978-2-35158-098-1; e-ISBN: 978-2-35158-099-8); *Eds. C. Miao, G. Ye and H. Chen*
PRO 72: First International Conference on 'Advances in Chemically-Activated Materials—CAM'2010' (2010), 264 pp., ISBN: 978-2-35158-101-8; e-ISBN: 978-2-35158-115-5; *Eds. Caijun Shi and Xiaodong Shen*
PRO 73: 2nd International Conference on 'Waste Engineering and Management—ICWEM 2010' (2010), 894 pp., ISBN: 978-2-35158-102-5; e-ISBN: 978-2-35158-103-2, *Eds. J. Zh. Xiao, Y. Zhang, M. S. Cheung and R. Chu*
PRO 74: International RILEM Conference on 'Use of Superabsorbent Polymers and Other New Additives in Concrete' (2010) 374 pp., ISBN: 978-2-35158-104-9; e-ISBN: 978-2-35158-105-6; *Eds. O.M. Jensen, M.T. Hasholt, and S. Laustsen*
PRO 75: International Conference on 'Material Science—2nd ICTRC—Textile Reinforced Concrete—Theme 1' (2010) 436 pp., ISBN: 978-2-35158-106-3; e-ISBN: 978-2-35158-107-0; *Ed. W. Brameshuber*

PRO 76: International Conference on 'Material Science—HetMat—Modelling of Heterogeneous Materials—Theme 2' (2010) 255 pp., ISBN: 978-2-35158-108-7; e-ISBN: 978-2-35158-109-4; *Ed. W. Brameshuber*
PRO 77: International Conference on 'Material Science—AdIPoC—Additions Improving Properties of Concrete—Theme 3' (2010) 459 pp., ISBN: 978-2-35158-110-0; e-ISBN: 978-2-35158-111-7; *Ed. W. Brameshuber*
PRO 78: 2nd Historic Mortars Conference and RILEM TC 203-RHM Final Workshop—HMC2010 (2010) 1416 pp., e-ISBN: 978-2-35158-112-4; *Eds. J. Válek, C. Groot and J. J. Hughes*
PRO 79: International RILEM Conference on Advances in Construction Materials Through Science and Engineering (2011) 213 pp., ISBN: 978-2-35158-116-2, e-ISBN: 978-2-35158-117-9; *Eds. Christopher Leung and K.T. Wan*
PRO 80: 2nd International RILEM Conference on Concrete Spalling due to Fire Exposure (2011) 453 pp., ISBN: 978-2-35158-118-6; e-ISBN: 978-2-35158-119-3; *Eds. E.A.B. Koenders and F. Dehn*
PRO 81: 2nd International RILEM Conference on Strain Hardening Cementitious Composites (SHCC2-Rio) (2011) 451 pp., ISBN: 978-2-35158-120-9; e-ISBN: 978-2-35158-121-6; *Eds. R.D. Toledo Filho, F.A. Silva, E.A.B. Koenders and E.M.R. Fairbairn*
PRO 82: 2nd International RILEM Conference on Progress of Recycling in the Built Environment (2011) 507 pp., e-ISBN: 978-2-35158-122-3; *Eds. V.M. John, E. Vazquez, S.C. Angulo and C. Ulsen*
PRO 83: 2nd International Conference on Microstructural-related Durability of Cementitious Composites (2012) 250 pp., ISBN: 978-2-35158-129-2; e-ISBN: 978-2-35158-123-0; *Eds. G. Ye, K. van Breugel, W. Sun and C. Miao*
PRO 84: CONSEC13—Seventh International Conference on Concrete under Severe Conditions—Environment and Loading (2013) 1930 pp., ISBN: 978-2-35158-124-7; e-ISBN: 978-2- 35158-134-6; *Eds. Z.J. Li, W. Sun, C.W. Miao, K. Sakai, O.E. Gjorv and N. Banthia*
PRO 85: RILEM-JCI International Workshop on Crack Control of Mass Concrete and Related issues concerning Early-Age of Concrete Structures—ConCrack 3—Control of Cracking in Concrete Structures 3 (2012) 237 pp., ISBN: 978-2-35158-125-4; e-ISBN: 978-2-35158-126-1; *Eds. F. Toutlemonde and J.-M. Torrenti*
PRO 86: International Symposium on Life Cycle Assessment and Construction (2012) 414 pp., ISBN: 978-2-35158-127-8, e-ISBN: 978-2-35158-128-5; *Eds. A. Ventura and C. de la Roche*
PRO 87: UHPFRC 2013—RILEM-fib-AFGC International Symposium on Ultra-High Performance Fibre-Reinforced Concrete (2013), ISBN: 978-2-35158-130-8, e-ISBN: 978-2-35158-131-5; *Eds. F. Toutlemonde*
PRO 88: 8th RILEM International Symposium on Fibre Reinforced Concrete (2012) 344 pp., ISBN: 978-2-35158-132-2; e-ISBN: 978-2-35158-133-9; *Eds. Joaquim A.O. Barros*
PRO 89: RILEM International workshop on performance-based specification and control of concrete durability (2014) 678 pp., ISBN: 978-2-35158-135-3; e-ISBN: 978-2-35158-136-0; *Eds. D. Bjegović, H. Beushausen and M. Serdar*

PRO 90: 7th RILEM International Conference on Self-Compacting Concrete and of the 1st RILEM International Conference on Rheology and Processing of Construction Materials (2013) 396 pp., ISBN: 978-2-35158-137-7; e-ISBN: 978-2-35158-138-4; *Eds. Nicolas Roussel and Hela Bessaies-Bey*

PRO 91: CONMOD 2014—RILEM International Symposium on Concrete Modelling (2014), ISBN: 978-2-35158-139-1; e-ISBN: 978-2-35158-140-7; *Eds. Kefei Li, Peiyu Yan and Rongwei Yang*

PRO 92: CAM 2014—2nd International Conference on advances in chemically-activated materials (2014) 392 pp., ISBN: 978-2-35158-141-4; e-ISBN: 978-2-35158-142-1; *Eds. Caijun Shi and Xiadong Shen*

PRO 93: SCC 2014—3rd International Symposium on Design, Performance and Use of Self-Consolidating Concrete (2014) 438 pp., ISBN: 978-2-35158-143-8; e-ISBN: 978-2-35158-144-5; *Eds. Caijun Shi, Zhihua Ou and Kamal H. Khayat*

PRO 94 (online version): HPFRCC-7—7th RILEM conference on High performance fiber reinforced cement composites (2015), e-ISBN: 978-2-35158-146-9; *Eds. H.W. Reinhardt, G.J. Parra-Montesinos and H. Garrecht*

PRO 95: International RILEM Conference on Application of superabsorbent polymers and other new admixtures in concrete construction (2014), ISBN: 978-2-35158-147-6; e-ISBN: 978-2-35158-148-3; *Eds. Viktor Mechtcherine and Christof Schroefl*

PRO 96 (online version): XIII DBMC: XIII International Conference on Durability of Building Materials and Components (2015), e-ISBN: 978-2-35158-149-0; *Eds. M. Quattrone and V.M. John*

PRO 97: SHCC3—3rd International RILEM Conference on Strain Hardening Cementitious Composites (2014), ISBN: 978-2-35158-150-6; e-ISBN: 978-2-35158-151-3; *Eds. E. Schlangen, M.G. Sierra Beltran, M. Lukovic and G. Ye*

PRO 98: FERRO-11—11th International Symposium on Ferrocement and 3rd ICTRC—International Conference on Textile Reinforced Concrete (2015), ISBN: 978-2-35158-152-0; e-ISBN: 978-2-35158-153-7; *Ed. W. Brameshuber*

PRO 99 (online version): ICBBM 2015—1st International Conference on Bio-Based Building Materials (2015), e-ISBN: 978-2-35158-154-4; *Eds. S. Amziane and M. Sonebi*

PRO 100: SCC16—RILEM Self-Consolidating Concrete Conference (2016), ISBN: 978-2-35158-156-8; e-ISBN: 978-2-35158-157-5; *Ed. Kamal H. Kayat*

PRO 101 (online version): III Progress of Recycling in the Built Environment (2015), e-ISBN: 978-2-35158-158-2; *Eds I. Martins, C. Ulsen and S. C. Angulo*

PRO 102 (online version): RILEM Conference on Microorganisms-Cementitious Materials Interactions (2016), e-ISBN: 978-2-35158-160-5; *Eds. Alexandra Bertron, Henk Jonkers and Virginie Wiktor*

PRO 103 (online version): ACESC'16—Advances in Civil Engineering and Sustainable Construction (2016), e-ISBN: 978-2-35158-161-2; *Eds. T.Ch. Madhavi, G. Prabhakar, Santhosh Ram and P.M. Rameshwaran*

PRO 104 (online version): SSCS'2015—Numerical Modeling—Strategies for Sustainable Concrete Structures (2015), e-ISBN: 978-2-35158-162-9

PRO 105: 1st International Conference on UHPC Materials and Structures (2016), ISBN: 978-2-35158-164-3; e-ISBN: 978-2-35158-165-0
PRO 106: AFGC-ACI-fib-RILEM International Conference on Ultra-High-Performance Fibre-Reinforced Concrete—UHPFRC 2017 (2017), ISBN: 978-2-35158-166-7; e-ISBN: 978-2-35158-167-4; *Eds. François Toutlemonde and Jacques Resplendino*
PRO 107 (online version): XIV DBMC—14th International Conference on Durability of Building Materials and Components (2017), e-ISBN: 978-2-35158-159-9; *Eds. Geert De Schutter, Nele De Belie, Arnold Janssens and Nathan Van Den Bossche*
PRO 108: MSSCE 2016—Innovation of Teaching in Materials and Structures (2016), ISBN: 978-2-35158-178-0; e-ISBN: 978-2-35158-179-7; *Ed. Per Goltermann*
PRO 109 (2 volumes): MSSCE 2016—Service Life of Cement-Based Materials and Structures (2016), ISBN Vol. 1: 978-2-35158-170-4; Vol. 2: 978-2-35158-171-4; Set Vol. 1&2: 978-2-35158-172-8; e-ISBN : 978-2-35158-173-5; *Eds. Miguel Azenha, Ivan Gabrijel, Dirk Schlicke, Terje Kanstad and Ole Mejlhede Jensen*
PRO 110: MSSCE 2016—Historical Masonry (2016), ISBN: 978-2-35158-178-0; e-ISBN: 978-2-35158-179-7; *Eds. Inge Rörig-Dalgaard and Ioannis Ioannou*
PRO 111: MSSCE 2016—Electrochemistry in Civil Engineering (2016); ISBN: 978-2-35158-176-6; e-ISBN: 978-2-35158-177-3; *Ed. Lisbeth M. Ottosen*
PRO 112: MSSCE 2016—Moisture in Materials and Structures (2016), ISBN: 978-2-35158-178-0; e-ISBN: 978-2-35158-179-7; *Eds. Kurt Kielsgaard Hansen, Carsten Rode and Lars-Olof Nilsson*
PRO 113: MSSCE 2016—Concrete with Supplementary Cementitious Materials (2016), ISBN: 978-2-35158-178-0; e-ISBN: 978-2-35158-179-7; *Eds. Ole Mejlhede Jensen, Konstantin Kovler and Nele De Belie*
PRO 114: MSSCE 2016—Frost Action in Concrete (2016), ISBN: 978-2-35158-182-7; e-ISBN: 978-2-35158-183-4; *Eds. Marianne Tange Hasholt, Katja Fridh and R. Doug Hooton*
PRO 115: MSSCE 2016—Fresh Concrete (2016), ISBN: 978-2-35158-184-1; e-ISBN: 978-2-35158-185-8; *Eds. Lars N. Thrane, Claus Pade, Oldrich Svec and Nicolas Roussel*
PRO 116: BEFIB 2016—9th RILEM International Symposium on Fiber Reinforced Concrete (2016), ISBN: 978-2-35158-187-2; e-ISBN: 978-2-35158-186-5; *Eds. N. Banthia, M. di Prisco and S. Soleimani-Dashtaki*
PRO 117: 3rd International RILEM Conference on Microstructure Related Durability of Cementitious Composites (2016), ISBN: 978-2-35158-188-9; e-ISBN: 978-2-35158-189-6; *Eds. Changwen Miao, Wei Sun, Jiaping Liu, Huisu Chen, Guang Ye and Klaas van Breugel*
PRO 118 (4 volumes): International Conference on Advances in Construction Materials and Systems (2017), ISBN Set: 978-2-35158-190-2; Vol. 1: 978-2-35158-193-3; Vol. 2: 978-2-35158-194-0; Vol. 3: ISBN:978-2-35158-195-7; Vol. 4: ISBN:978-2-35158-196-4; e-ISBN: 978-2-35158-191-9; *Ed. Manu Santhanam*

PRO 119 (online version): ICBBM 2017—Second International RILEM Conference on Bio-based Building Materials, (2017), e-ISBN: 978-2-35158-192-6; *Ed. Sofiane Amziane*
PRO 120 (2 volumes): EAC-02—2nd International RILEM/COST Conference on Early Age Cracking and Serviceability in Cement-based Materials and Structures, (2017), Vol. 1: 978-2-35158-199-5, Vol. 2: 978-2-35158-200-8, Set: 978-2-35158-197-1, e-ISBN: 978-2-35158-198-8; *Eds. Stéphanie Staquet and Dimitrios Aggelis*
PRO 121 (2 volumes): SynerCrete18: Interdisciplinary Approaches for Cement-based Materials and Structural Concrete: Synergizing Expertise and Bridging Scales of Space and Time, (2018), Set: 978-2-35158-202-2, Vol.1: 978-2-35158-211-4, Vol.2: 978-2-35158-212-1, e-ISBN: 978-2-35158-203-9; *Eds. Miguel Azenha, Dirk Schlicke, Farid Benboudjema, Agnieszka Knoppik*
PRO 122: SCC'2018 China—Fourth International Symposium on Design, Performance and Use of Self-Consolidating Concrete, (2018), ISBN: 978-2-35158-204-6, e-ISBN: 978-2-35158-205-3; *Eds. C. Shi, Z. Zhang, K. H. Khayat*
PRO 123: Final Conference of RILEM TC 253-MCI: Microorganisms-Cementitious Materials Interactions (2018), Set: 978-2-35158-207-7, Vol.1: 978-2-35158-209-1, Vol.2: 978-2-35158-210-7, e-ISBN: 978-2-35158-206-0; *Ed. Alexandra Bertron*
PRO 124 (online version): Fourth International Conference Progress of Recycling in the Built Environment (2018), e-ISBN: 978-2-35158-208-4; *Eds. Isabel M. Martins, Carina Ulsen, Yury Villagran*
PRO 125 (online version): SLD4—4th International Conference on Service Life Design for Infrastructures (2018), e-ISBN: 978-2-35158-213-8; *Eds. Guang Ye, Yong Yuan, Claudia Romero Rodriguez, Hongzhi Zhang, Branko Savija*
PRO 126: Workshop on Concrete Modelling and Material Behaviour in honor of Professor Klaas van Breugel (2018), ISBN: 978-2-35158-214-5, e-ISBN: 978-2-35158-215-2; *Ed. Guang Ye*
PRO 127 (online version): CONMOD2018—Symposium on Concrete Modelling (2018), e-ISBN: 978-2-35158-216-9; *Eds. Erik Schlangen, Geert de Schutter, Branko Savija, Hongzhi Zhang, Claudia Romero Rodriguez*
PRO 128: SMSS2019—International Conference on Sustainable Materials, Systems and Structures (2019), ISBN: 978-2-35158-217-6, e-ISBN: 978-2-35158-218-3
PRO 129: 2nd International Conference on UHPC Materials and Structures (UHPC2018-China), ISBN: 978-2-35158-219-0, e-ISBN: 978-2-35158-220-6
PRO 130: 5th Historic Mortars Conference (2019), ISBN: 978-2-35158-221-3, e-ISBN: 978-2-35158-222-0; *Eds. José Ignacio Álvarez, José María Fernández, Íñigo Navarro, Adrián Durán, Rafael Sirera*
PRO 131 (online version): 3rd International Conference on Bio-Based Building Materials (ICBBM2019), e-ISBN: 978-2-35158-229-9; *Eds. Mohammed Sonebi, Sofiane Amziane, Jonathan Page*
PRO 132: IRWRMC'18—International RILEM Workshop on Rheological Measurements of Cement-based Materials (2018), ISBN: 978-2-35158-230-5, e-ISBN: 978-2-35158-231-2; *Eds. Chafika Djelal, Yannick Vanhove*

PRO 133 (online version): CO2STO2019—International Workshop CO2 Storage in Concrete (2019), e-ISBN: 978-2-35158-232-9; *Eds. Assia Djerbi, Othman Omikrine-Metalssi, Teddy Fen-Chong*
PRO 134: 3rd ACF/HNU International Conference on UHPC Materials and Structures—UHPC'2020, ISBN: 978-2-35158-233-6, e-ISBN: 978-2-35158-234-3; *Eds. Caijun Shi and Jiaping Liu*
PRO 135: Fourth International Conference on Chemically Activated Materials (CAM2021), ISBN: 978-2-35158-235-0, e-ISBN: 978-2-35158-236-7; *Eds. Caijun Shi and Xiang Hu*
PRO 136: Lime Based Materials for Repairing Historic Structures (2022), ISBN: 978-2-35158-240-4, e-ISBN: 978-2-35158-237-4; *Eds. Ioanna Papayianni and Jan Válek*
PRO 137 (online version): V International Conference Progress of Recycling in the Built Environment (2023), e-ISBN: 978-2-35158-238-1; *Eds. Ines Döring, Yury Villagrán Zaccardi, Anette Müller*

RILEM Reports (REP)

Report 19: Considerations for Use in Managing the Aging of Nuclear Power Plant Concrete Structures (ISBN: 2-912143-07-1); *Ed. D. J. Naus*
Report 20: Engineering and Transport Properties of the Interfacial Transition Zone in Cementitious Composites (ISBN: 2-912143-08-X); *Eds. M. G. Alexander, G. Arliguie, G. Ballivy, A. Bentur and J. Marchand*
Report 21: Durability of Building Sealants (ISBN: 2-912143-12-8); *Ed. A. T. Wolf*
Report 22: Sustainable Raw Materials—Construction and Demolition Waste (ISBN: 2-912143-17-9); *Eds. C. F. Hendriks and H. S. Pietersen*
Report 23: Self-Compacting Concrete state-of-the-art report (ISBN: 2-912143-23-3); *Eds. Å. Skarendahl and Ö. Petersson*
Report 24: Workability and Rheology of Fresh Concrete: Compendium of Tests (ISBN: 2-912143-32-2); *Eds. P. J. M. Bartos, M. Sonebi and A. K. Tamimi*
Report 25: Early Age Cracking in Cementitious Systems (ISBN: 2-912143-33-0); *Ed. A. Bentur*
Report 26: Towards Sustainable Roofing (Joint Committee CIB/RILEM) (CD 07) (e-ISBN 978-2-912143-65-5); *Eds. Thomas W. Hutchinson and Keith Roberts*
Report 27: Condition Assessment of Roofs (Joint Committee CIB/RILEM) (CD 08) (e-ISBN 978-2-912143-66-2); *Ed. CIB W 83/RILEM TC166-RMS*
Report 28: Final report of RILEM TC 167-COM 'Characterisation of Old Mortars with Respect to Their Repair (ISBN: 978-2-912143-56-3); *Eds. C. Groot, G. Ashall and J. Hughes*
Report 29: Pavement Performance Prediction and Evaluation (PPPE): Interlaboratory Tests (e-ISBN: 2-912143-68-3); *Eds. M. Partl and H. Piber*

Report 30: Final Report of RILEM TC 198-URM 'Use of Recycled Materials' (ISBN 2-912143-82-9; e-ISBN: 2-912143-69-1); *Eds. Ch. F. Hendriks, G. M. T. Janssen and E. Vázquez*

Report 31: Final Report of RILEM TC 185-ATC 'Advanced testing of cement-based materials during setting and hardening' (ISBN: 2-912143-81-0; e-ISBN: 2-912143-70-5); *Eds. H. W. Reinhardt and C. U. Grosse*

Report 32: Probabilistic Assessment of Existing Structures. A JCSS publication (ISBN 2-912143-24-1); *Ed. D. Diamantidis*

Report 33: State-of-the-Art Report of RILEM Technical Committee TC 184-IFE 'Industrial Floors' (ISBN 2-35158-006-0); *Ed. P. Seidler*

Report 34: Report of RILEM Technical Committee TC 147-FMB 'Fracture mechanics applications to anchorage and bond' Tension of Reinforced Concrete Prisms—Round Robin Analysis and Tests on Bond (e-ISBN 2-912143-91-8); *Eds. L. Elfgren and K. Noghabai*

Report 35: Final Report of RILEM Technical Committee TC 188-CSC 'Casting of Self Compacting Concrete' (ISBN 2-35158-001-X; e-ISBN: 2-912143-98-5); *Eds. Å. Skarendahl and P. Billberg*

Report 36: State-of-the-Art Report of RILEM Technical Committee TC 201-TRC 'Textile Reinforced Concrete' (ISBN 2-912143-99-3); *Ed. W. Brameshuber*

Report 37: State-of-the-Art Report of RILEM Technical Committee TC 192-ECM 'Environment-conscious construction materials and systems' (ISBN: 978-2-35158-053-0) *Eds. N. Kashino, D. Van Gemert and K. Imamoto*

Report 38: State-of-the-Art Report of RILEM Technical Committee TC 205-DSC 'Durability of Self-Compacting Concrete' (ISBN: 978-2-35158-048-6); *Eds. G. De Schutter and K. Audenaert*

Report 39: Final Report of RILEM Technical Committee TC 187-SOC 'Experimental determination of the stress-crack opening curve for concrete in tension' (ISBN 978-2-35158-049-3); *Ed. J. Planas*

Report 40: State-of-the-Art Report of RILEM Technical Committee TC 189-NEC 'Non-Destructive Evaluation of the Penetrability and Thickness of the Concrete Cover' (ISBN 978-2-35158-054-7); *Eds. R. Torrent and L. Fernández Luco*

Report 41: State-of-the-Art Report of RILEM Technical Committee TC 196-ICC 'Internal Curing of Concrete' (ISBN 978-2-35158-009-7); *Eds. K. Kovler and O. M. Jensen*

Report 42: 'Acoustic Emission and Related Non-destructive Evaluation Techniques for Crack Detection and Damage Evaluation in Concrete'—Final Report of RILEM Technical Committee 212-ACD (e-ISBN: 978-2-35158-100-1); *Ed. M. Ohtsu*

Report 45: Repair Mortars for Historic Masonry—State-of-the-Art Report of RILEM Technical Committee TC 203-RHM (e-ISBN: 978-2-35158-163-6); *Eds. Paul Maurenbrecher and Caspar Groot*

Report 46: Surface delamination of concrete industrial floors and other durability related aspects guide—Report of RILEM Technical Committee TC 268-SIF (e-ISBN: 978-2-35158-201-5); *Ed. Valerie Pollet*

Contents

Introduction to Active Control of Properties of Fresh and Hardening Cementitious Materials

Geert De Schutter and Jay Sanjayan

Abstract This chapter gives a general introduction to the State-of-the-Art Report on the Active Control of Properties of Fresh and Hardening Cementitious Materials (ACP). Active control of properties of cementitious materials refers to the action of controlling some specific properties (e.g. rheology, stiffening, …) by applying an external trigger signal that induces some response in the cementitious material, with a modification of properties or behaviour as a result. An overview is given of some historical active control methodologies. The potential trigger signals and responses are briefly discussed. Finally, the structure of the State-of-the-Art Report is outlined.

Keywords Active control · Cementitious materials · Fresh properties · Hardening properties

1 Introduction

Active Control of Concrete Properties is a recently identified sub-domain within the field of concrete technology. A first (p)review paper was published in 2018 in Materials and Structures [1]. The text of this previously published Open Access paper is largely taken as the basis for current introduction chapter.

Concrete properties largely depend on mix design and processing, leaving only limited options to actively modify concrete properties after mixing, during processing and casting. The new concept of Active Control of concrete properties is based on the application of external signals to trigger an intended response in the material. Some current practices in concrete industry could be considered as active control, e.g. mechanical vibration, steam curing, cooling of mass concrete. More advanced

G. De Schutter (✉)
Magnel-Vandepitte Laboratory, Department of Structural Engineering and Building Materials, Ghent University, Ghent, Belgium
e-mail: Geert.DeSchutter@ugent.be

J. Sanjayan
Swinburne University of Technology, Melbourne, Australia

G. De Schutter and J. Sanjayan (eds.), *Active Control of Properties of Fresh and Hardening Cementitious Materials*, RILEM State-of-the-Art Reports 50,
https://doi.org/10.1007/978-3-032-30340-0_1

active control mechanisms can be inspired by methodologies in other fields, e.g. based on hydrogels and other functional polymers.

A specific approach recently studied in more detail focuses on active rheology control (ARC) [2]. Active rheology control of cementitious materials can be based on magneto-responsive mineral particles, and/or triggerable functional polymers. However, several challenges remain, like the stability and functioning of the responsive material in a cementitious environment, the applicability of the control signal in a cementitious material, and the cost, logistics and safety of a control system on a construction site or in precast industry. Finding solutions to these challenges will lead to marvellous opportunities in general, and for concrete 3D printing more particularly.

Besides rheology, active control can also be directed towards other properties of concrete. The main interest seems to go to the processing stage (pumping, casting, 3D printing…), but other aspects besides processing are also of interest, e.g. control of setting, curing and shrinkage during the hardening stage of the cementitious material. A whole new world of possibilities can be provided by active control features, moving beyond the current status of passively relying on the behaviour of the cementitious materials as mainly determined by the mix design, the mixing process, and the environmental conditions during processing and hardening. An introduction to this new world is presented in this RILEM state-of-the-art report.

2 Active Control: Definition

Active control of properties of cementitious materials refers to the action of controlling some specific properties (e.g. rheology, stiffening, …) by applying an external trigger signal that induces some response in the cementitious material, with a modification of properties or behaviour as a result. In the ideal active control process, the intervention is based on observations and/or measurements (quantified), giving the necessary information to take a decision regarding required intervention. A feed-back loop between measurement and signal seems an optimal situation.

Active control thus involves an additional step compared to passively controlling the cementitious properties by careful mix design and optimal mixing process. Control of properties in literature most often refers to the concept of optimizing relevant mix design parameters (including e.g. water/cement-ratio, cement content) and/or adding specific admixtures (e.g. superplasticizers, air-entraining agents) or additions (e.g. silica fume, metakaolin) in view of reaching desired properties. After mixing, the performance of the cementitious material cannot be further adjusted, and the workers have no significant means to alter the performance in real time. This is different in the case of active control, which involves triggering a desired response of the cementitious material on demand, within the further processing steps after the mixing stage. The response can be obtained by actions of classical constituents (e.g. cement or fly ash particles), or by specially designed responsive admixtures or (nano-)particles. A detailed literature overview of potential options is described in the different chapters of this state-of-the-art report.

3 Historical Active Control Methodologies

Until recently, it was never explicitly called active control. Nevertheless, some current and historical methodologies can be considered as active control *avant la lettre*. An overview is given hereafter, largely based on the open access publication [1] published by Springer.

In current concrete practice, active interventions are manifold, although mostly not in the sense of actively modifying material behaviour by triggering a response of mixed-in special components. Nevertheless, an overview of current active intervention methods (as also schematically summarized in Table 1) is a good starting point to learn from, and to push the limit of our current abilities to control fresh and hardened concrete properties.

3.1 Mechanical Intervention

The direct contact between (aggregate) particles leads to high friction, obstructing the flow. In fresh state, an obvious active mechanical intervention is given by the vibration process, using poker vibration or other techniques. By means of a poker, a vibration with a certain frequency and amplitude is (in this case internally) imposed to the concrete. This “signal” has a certain range of action, triggering the (solid) particles in the cementitious material to respond (vibrate). Consequently, the yield stress drops to zero within the affected zone [3], and the material liquefies and fills the formwork in a better way (i.e. compacts). While a poker requires an invasive action (‘internal vibration’), other vibration methods are available operating from the surface of the concrete, e.g. formwork vibrators. External vibrators on pumping pipes are sometimes used at regular distances in case of pumping of concrete for tall buildings, several hundreds of meters upwards.

Vibration of fresh concrete can thus be seen as the active application of an external signal (mechanical vibration with a certain frequency and amplitude), triggering a response from some constituents in the bulk of the concrete (in this case vibration of solid particles), changing the rheological properties of the fresh concrete (reducing the yield stress to zero). In this way, it could be said that active rheology control of fresh concrete has been already performed in concrete industry since the 1930s due to the introduction of poker vibrators. However, active rheology control can go much further, as will be illustrated further in this report.

Table 1 Overview of current practices of active intervention

Type of intervention	"Signal"	"Responsive" constituents	Effect
Mechanical	Vibration (frequency, amplitude, range)	Solid particles (vibration)	Elimination of yield stress (rheology control)
Chemical	Chemical trigger, e.g. retarder, accelerator, CO_2 …	Cement particles, molecules in pore solution …	Retardation or acceleration of hydration reaction … (setting and hardening control)
Thermal	Temperature field	All constituents (through their thermal properties), cement particles (activation energy)	Control of temperature gradients, control of reaction rate
Hygral	Moisture	Pore system (capillary pressure, diffusion …), cement particles (reactivity)	Control of shrinkage (cracking), control of hydration process
Pressure	Reduced or increased pressure, or ultrasound	Air	Control of compaction, contribution to accelerated curing, dispersion
Magnetic	Magnetic field	Magnetizable materials (MNPs, steel strips or fibres…), water	Set-on-demand, micro-vibration, effect of magnetized water on hydration, healing on demand …
Electric	Electric current or potential	Dipolar molecules, ions	Control of reaction rate (via heating due to resistivity), formwork release, improved pumping (via polarization)
Microwave	High frequency electric signal	Dipolar elements	Control of reaction rate (via heating due to internal dipolar vibration)

3.2 *Chemical Intervention*

An active chemical intervention is often applied in case of sprayed concrete, introducing accelerating admixtures in the spraying head. The added accelerating admixture triggers the cement particles, speeding up the hydration process in a chemical way, leading to faster setting (stiffening) and hardening of the sprayed concrete. Similar but even more complex chemical interventions can be applied when at the end of a working day the concrete within a long pumping pipe (e.g. for the construction of a tall tower) is removed for pipe cleaning. The concrete flowing back in a truck mixer can be retarded for a long night by adding a chemical retarder. The next morning, while resuming pumping operations, a chemical accelerator can be added, counteracting the effect of the retarder. This kind of active combination of retarders and accelerators has e.g. been studied by [4].

The active intervention in this case is giving a chemical signal to the cement particles, leading to a control of the hydration rate. After addition of the accelerator the triggered response of the cement particles is irreversible. However, the triggered response of the added retarder can be reversed by later addition of an accelerator. This kind of active intervention is limited to certain moments in the process, e.g. when the concrete is flowing through the spray head in case of sprayed concrete, or when the concrete is in the truck mixer in the case of re-use of concrete in a pumping pipe. The control is no longer possible at other times, and it is also not possible to adjust the control levels once the admixtures have been added.

Another example of active chemical intervention is the use of carbon dioxide for the accelerated curing of cementitious materials, often in combination with the application of a higher temperature. This active chemical intervention has already been reported in 1972 by [5], but has been given renewed attention in recent years because of environmental issues, as illustrated by Monkman et al. [6], studying the addition of carbon dioxide gas during mixing of the concrete for the production of blocks, and by Nielsen et al. [7], studying carbonate-bonded construction materials from alkaline residues.

3.3 Thermal Intervention

The effect of temperature on the hydration process of cement is well-known and can be calculated by means of the Arrhenius Equation [8], with the activation energy as most important parameter. This knowledge is applied in industry by means of heat curing, e.g. steam cycles in precast industry [9, 10]. This involves active intervention, applying a thermal signal to the hardening concrete with a view to accelerating the hydration process and achieving faster setting and strength development (at the cost of reducing somewhat the final strength [11]). The applied thermal field is triggering the cement particles (and/or supplementary cementitious materials) that are responding according to their activation energy, resulting in an acceleration of the hydration process. Steam curing (and heat curing more generally) can be considered as a type of active setting and hardening control.

Another type of active control involving a thermal field is given by cooling measures to mitigate thermal cracking in hardening massive concrete elements [12]. In embedded cooling pipes, a cooling agent is circulated through the massive element, imposing lower temperatures inside the hardening concrete [13, 14]. All constituents of the concrete respond to the applied thermal signal (lower temperature in embedded cooling pipes) following the physical laws of thermal diffusion, as described by Fourier's Equation [15]. Important parameters in this process are the specific heat and the thermal diffusivity of the hardening concrete, as well as the heat production due to the exothermal hydration process [16], also responsive to temperature following the previously mentioned Arrhenius Equation.

3.4 Hygral Intervention

Freshly cast concrete needs to be protected against premature drying, in view of mitigating shrinkage cracking and creating optimal conditions for the hydration process. For onsite casting, protection against premature drying during the first days is for many elements achieved in a passive way by the formworks. After striking formwork, or for those elements not protected by formwork (e.g. slabs), curing can be performed in an active way by water spraying or other techniques. The active intervention providing sufficient moisture near the concrete surface is interacting with the pore system, based on capillary forces and diffusion processes. As a result, the concrete is maintaining sufficient moisture conditions, mitigating or reducing early age shrinkage cracking, and improving the hydration process [17].

Modern concepts of internal curing by means of porous aggregate particles or super absorbent polymers (SAP) [18] cannot be considered as active interventions, as the water contained in those materials is only released internally because of the self-desiccation linked to the hydration process. In future, however, active intervention could be envisioned by making SAPs or similar products to release their water content on demand by providing an external trigger. The triggered release of agents contained in hydrogel-type polymers is a known concept in other fields already now.

3.5 Intervention by Pressure

Concrete also includes air, showing a potential effect on the rheological properties depending on the size of air bubbles and the applied shear stresses [19]. During processing, air can influence the rheological behaviour, and vice versa, the processing can influence the air system of the concrete [20]. During mixing, pressures in the mixing pan can be reduced, resulting in vacuum mixing [21]. As a result, the air void system of the fresh (and later of hardened) concrete is influenced, resulting in reduced porosity and increased mechanical and durability performance. Modifying the air content in fresh concrete also influences the rheological behaviour [19]. Others made use of frequently changing pressure waves by means of an inline ultrasound application [22]. A process referred to as cavitational shear disperses cement grains better and thereby reduces the setting time and admixture use. This pre-cast method allows to speed up stiffening and is preferentially applied to neat cement paste, before mixing it as concrete. Vacuum conditions can also be applied during casting, resulting in vacuum compaction of the fresh concrete [23]. While curing, air pressures could be increased, possibly combined with increased temperature and/or increased CO_2-content (see also higher), in view of accelerated hardening of the concrete element [24]. External pressure control can thus trigger changes in air content and pore structure, in view of achieving improved material performance.

3.6 Magnetic Intervention

Magnetic fields can be relatively easily applied in industry and can be an element of control during a production process. A magnetic field could be applied to magnetize water before its addition to cementitious materials. A magnetic field changes the physical properties of water, including e.g. specific heat and evaporation rate [25]. Some researchers claim that the use of magnetized water in cementitious materials influences fresh and hardened properties, e.g. Reddy et al. in [26]. They observed that the use of magnetized water in concrete resulted in a very significant increase in concrete strength, probably due to the higher surface area of magnetized water. Su et al. [27] also observed a strength increase when using magnetized water in mortar and concrete containing fly ash. Nevertheless, the use of magnetized water is still open to significant debate, as it is not clear how water could maintain its magnetization long enough after removal of the magnetic field. It has to be further studied whether the reported improvement of properties really is the effect of previous magnetization of water, or rather the effect of other experimental influences that have not been clearly identified.

Soto-Bernal et al. [28] studied the effect of maintained static magnetic fields applied during hardening on the performance of cement paste. They concluded that more CSH is produced, with denser morphology and reduced porosity. It is conjectured that the magnetic field is influencing the hydration process by means of a restructuring process at molecular scale, resulting in microstructural improvement and increased mechanical performance.

The application of magnetic fields to control rheology of cementitious materials containing magnetic particles in real time has been studied by Nair and Ferron [29]. They considered two types of carbonyl iron powder, added to Portland cement paste (water cement ratio 0.4) in dosages up to 10% by mass of cement (corresponding to 4% in volume). Magnetic fields with strengths up to 1.0 T have been applied while performing rheological tests on the fresh samples. It is experimentally shown that the rheological behaviour of the paste containing magnetic particles can be actively altered by applying the magnetic field. The effect depends on the field strength as well as on the dosage of magnetic particles. On the other hand, the authors conclude that the magnetic field does not seem to have a noticeable effect on the morphology and formation of early age hydration product and does not influence the final compressive strength. This seems to be in contradiction with the previously mentioned conclusions obtained by Soto-Bernal et al. [28], showing that further research is needed to elucidate the effect of magnetic fields on cement hydration and microstructure development. Disregarding the discussion on the effect of magnetic field on hydration, the research of Nair and Ferron [29] is a major step forward towards active rheology control of cementitious materials.

Abavisani et al. [30] also considered magnetic fields to study the possibility to align steel fibres (chips) according to a certain orientation. They conclude that the application of a magnetic field on the fresh steel chip-reinforced concrete facilitates the compaction process and leads to an increased compressive strength depending on the test direction. Following the optimum direction, the compressive strength increases by more than 17%. Particularly interesting regarding rheology control is their observation that the vibration of the steel chips in the alternating magnetic field can be considered as a way to obtain *self-compactor elements* for concrete.

The same group of authors further studied the effect of alternating magnetic fields and alternating current on the behaviour of reinforced concrete columns [31], considering magnetic fields up to 0.5 T with frequency of 50 Hz, and AC current up to 36 A. The alternating magnetic field facilitated the compaction of fresh concrete in the reinforced columns due to the triggered vibration of the reinforcing bars.

Applying the magnetic field on the hardened concrete seems to lead to contradictory results, with sometimes an increase [30] and sometimes a decrease [31] in performance. The authors state that the application of magneto-electric fields can be the basis for *real-time compressive behaviour controlling* of reinforced concrete elements, even enabling a control of failure mode. However, their claims are based on contradictory results and are not fundamentally supported by a clear mechanism explaining the potential influence of the (electro-)magnetic field on the material and structural properties. While the principle of active control of concrete properties by magnetic signals has been set forward, it is not fully clear which mechanism could be the basis for the explanation of the reported effect of magnetization on hydration and microstructure development. This will need further research attention, with more rigorous test protocols to fully exclude potential influences of other parameters.

On the other hand, it seems clear that mixed-in magnetic particles or magnetizable materials like steel chips as well as preplaced steel rebars are responding to the magnetic signal, with effects on rheology, compactability and even mechanical performance. This opens a window for new developments and applications, possibly involving new types of magnetizable materials [32] or new fields of application like on-demand healing [33, 34]. In some applications, the effect of the magnetic field is triggering its response via another physical mechanism like (micro-)vibration or temperature increase.

3.7 Electric Intervention

An electric control signal (AC current) was mentioned in previous paragraph, where it was combined with the research on the magnetic control of the mechanical performance of hardened reinforced concrete columns. Other literature results are available on the application of electric fields on fresh or hardening concrete. As an example, Kovtun et al. [35] apply direct current for the curing of alkali-activated fly ash concrete. The current serves as an indirect way of heating, while the higher temperature achieved is leading to the accelerated hardening of the material. In their research, Kovtun et al. used a setup in which the direct current, provided through a controllable unit with maximum current 12 A and voltage up to 260 V, was continuously controlled in order to reach a curing temperature of 60 °C in the concrete, with an initial heating rate of 20 °C in order to avoid too fast heating, similar to what is done when applying steam cycles. While the external signal is electric, the physically controlling mechanisms behind the accelerated hardening is thermal energy. Nevertheless, the response of the alkali activated material to the electric trigger depends on the activator used, through its influence on the resistivity of the concrete. A similar principle has been followed by Zhang et al. [36, 37] as described in their patents concerning the control of setting time of alkali-activated concrete through the application of an electric field. The electric field only serves as a practical means to control the temperature in the cementitious material, depending on its resistivity.

A more innovative application of an electrical field can be found in the papers of Goudjil et al. [38, 39]. They impose an electric potential on a steel formwork panel, up to a level of 3 V for 10 minutes, in view of polarizing the concrete layer close to the formwork surface. The applied electric signal triggers a response in the material, resulting in a reduced adhesion between the hardening concrete and the formwork panel. The induced electro-osmotic effect is thus facilitating the formwork release once the concrete has sufficient strength to be self-supporting. In this way, the application of demoulding oils can be avoided.

Inspired by this electrically controlled formwork removal technique, Choi et al. [40] applied a similar polarization technique to improve the flow of concrete in pumping pipes. In the first pilot test, they externally applied an electromagnetic field around steel pumping pipes to polarize the material near the pipe surface and improve the formation of the lubrication layer. Although the electromagnetic field applied by Choi et al. indeed seemed to improve the pumping behaviour of the concrete, as experimentally verified by an ultrasound velocity profiling technique, it is important to notice that they did not add any specifically responding material like magnetic particles to the concrete. They merely relied on the response of the classical constituents present in the concrete. Although they mentioned the increased mobility of magnetized water and the results of the pilot test were positive, the fundamental physico-chemical mechanisms behind the success of this technique need further study and elucidation. Understanding this in detail might open the door to the development of new materials that could be mixed-in, strengthening its response triggered by the polarization signal and potentially enabling to better control the response to an intended level.

Another example of electric intervention is the technique of cathodic protection by means of applied current in view of protecting an existing structure from reinforcement corrosion [41]. This technique is becoming more popular in concrete practice and is a nice illustration of currently available active control mechanisms. The fundamental principle of its functioning can be understood by considering a Pourbaix diagram, explaining passive and active states of reinforcing steel in concrete [42].

3.8 Intervention with Microwaves

As an extension of electric signals, microwaves can also be mentioned in this overview. Microwaves are electromagnetic waves with high frequencies (300 MHz–300 GHz) and wavelengths ranging from one millimeter to one meter. A dielectric material placed in a high frequency electric field will heat up due to internal friction, as the induced vibration of the dipolar molecules is hindered by molecular attraction forces. As already reported in 1987 by Xuequan et al. [43], the promoted cement hydration can be attributed to the elevated temperature as a result of the exposure of the hardening concrete to microwaves. However, an additional effect seems to be the desiccation of free water prior to setting, resulting in decreased porosity. As a result, microwave curing of concrete yields very high early strength, while also providing slightly higher strength at 7 and 28 days [43, 44].

4 Potential Trigger Signals and Responses

Different types of trigger signals can be considered, in combination with different types of potential responses. An overview of different approaches in different disciplines is given hereafter, largely based on the open access publication [1] published by Springer.

Most common signals concern temperature and pH (or pH difference), or the application of an electric, magnetic, or electromagnetic field. Many other signals, however, have shown to be applicable in view of inducing a triggered response in functional materials, like e.g. pressure [45], redox-signal [46–51], carbon dioxide [48, 52, 53], light [50, 54], ultrasound [55], UV [56–58] …

Most stimuli-responsive polymers respond to one or more trigger signals. In case of dual or multi-responsive materials, the desired action depending on the given combination of trigger signals is an important point of attention. In case of two signals, as an example, it could be that the desired response is only activated in the presence of both stimuli, while another option could be the activation in case of any single stimulus. Different other logic combinations can be chosen, as illustrated in [59].

The applicability of the trigger signal depends on the type of material and further depends on practical considerations. While for some applications photonic signals are convenient and environmentally friendly, the responsive process (often chemical) is typically not showing a high efficiency [60]. For opaque materials like cementitious materials, light signals are not a good option, unless only a response at the surface is needed. Signals based on change in pH or on electrochemical responses are typically having higher efficiency. Electric or redox signals further depend on the resistivity or conductivity of the material and are thus not applicable in all circumstances. Heating and cooling seem easy trigger systems in principle but are not very practical in many cases.

Roy et al. [61] give an interesting overview of stimuli-responsive materials, with clear illustrations of their effects. Numerous types of responses, resulting in different effects can be listed. In general, control can be reversible or non-reversible, depending on the type of triggered response and the type of signal. A non-exhaustive list of different types of triggered responses is given hereafter:

- Triggering crosslinking resulting in gel formation, often by means of thermally induced sol-gel transition [62]
- Cleavage of polymer molecules, resulting in collapse of a hydrogel or a hyperbranched polymer, or influencing micellization [54, 55, 58]
- Controlling swelling/de-swelling and the release of an agent, due to changes in osmotic pressure by asymmetric distribution of ions in electric field [63], or by change of electronic charge surface potential [64]
- Changing the permeability of a responsive membrane, by closing or opening "chemical gates", e.g. by pH control [61]
- Sorption control of water vapour of e.g. electro-responsive polypyrole film [65]
- Control of solubility, often by changing temperature [66]
- Changing the nature of a complex molecule from hydrophilic to hydrophobic, also enabling control over the micellization process [62]
- Changing the self-assembly, e.g. CO_2 and redox-stimulated transformation of entangled worms to rodlike micelles [48]
- Changing the orientation of the polymers, e.g. by the application of an electric and/or magnetic field [67, 68]
- Controlling entanglement of polymer chains [69]
- Controlling aggregation/dispersion of particles in a medium [56]
- Control of trans-cis isomerization of polymers [57]
- Shape-morphing [70]
- Oscillatory shape change, e.g. oscillatory bending [70, 71]
- Crystallization/melting transition [72]
- Crystal anisotropic/isotropic transition [72]
- …

The listed responses can induce changes in numerous material properties or can lead to specific actions (like e.g. drug release). Some of the listed responses could have good potential for rheology control.

5 A New Concrete Technology Sub-Field

Triggered by the results obtained within the ERC Advanced Grant Project 'Smart-Cast', as published in an open access book [2] and in many journal papers, several groups worldwide started studying active control for different properties and/or different processes. This initiated a new sub-field withing concrete technology research, named Active Control.

To streamline the new terminology and methodology in support of the growing number of active control research projects, in 2023 a RILEM Technical committee was launched on this subject, entitled 'Active Control of Properties of Fresh and Hardening Concrete'. The main purpose of this TC was to create a State-of-the-art report on Active Control of Properties of Cementitious Materials, not only focussing on rheology but also on other properties in fresh state and during hardening. The committee work did not focus on active control in hardened state, but major hardened material characteristics are referred to as potentially influenced by active control in fresh and hardening state.

The members of the TC performed bibliographical research. No new research was initiated within the committee itself. As the draft chapters of this state-of-the-art report have been written in fall 2024 and spring 2025, literature and reports published before end of 2024 or at the latest early 2025 then have been considered in the review.

6 Structure of this Report

The focus of this State-of-the-art report on Active Control of Properties of Cementitious Materials is not only on rheology but also on other properties in fresh state and during hardening. The report does not focus on active control in hardened state, although major hardened material characteristics will be referred to as potentially influenced by active control in fresh and hardening state.

After the (Introduction to Active Control of Properties of Fresh and Hardening Cementitious Materials) and a (Glossary), this report consists of two major parts. First, the state-of-the-art report describes (Active Control of Rheology of Cementitious Materials), (Active Control of Setting of Cementitious Materials), and (Active Control of Concrete Shrinkage and Associated Cracking). Afterwards, the focus is on (Active Control of Concrete Pumping), (Active Control of Formwork Casting of Cementitious Materials), (Active Control of 3D Printing of Cementitious Materials), and (Active Control of Cementitious Slabs and Roads). Finally, Chap. 10 presents a Summary and Further Perspectives Regarding Active Control of Cementitious Materials.

References

1. De Schutter, G., Lesage, K.: Active control of properties of concrete—a (p) review. Mater. Struct. **51**, 123 (2018). https://doi.org/10.1617/s11527-018-1256-2
2. De Schutter, G., Lesage, K.: Active Rheology Control of Cementitious Materials, Taylor & Francis, London, UK. ISBN 978–1–0032-8946-3, pp. 298 (2023). https://doi.org/10.1201/9781003289463
3. Grampeix, G., Roussel, N., Dupoirier J.: Internal vibration and viscous concrete: application and prediction of the radius of action. In: Roussel, N., Bessaies-Bey, H. (eds.) In: Rheology

and Processing of Construction Materials, RILEM Proceedings PRO 90. RILEM Publications, pp. 123–130. ISBN 978-2-35158-137-7 (2013)
4. Lesage, K.: Interactions between cement and combined concrete admixtures. The influence on cement paste rheology. PhD thesis, joint PhD University of Leuven and Ghent University, Belgium (2014)
5. Klemm, W.A., Berger, R.L.: Accelerated curing of cementitious systems by carbon dioxide. Part I. Portland cement. Cem. Concr. Res. **2**, 567–576 (1972)
6. Monkman, S., MacDonald, M.: Carbon dioxide upcycling into industrially produced concrete blocks. Constr. Build. Mater. **124**, 127–132 (2016)
7. Nielsen, P., Baciocchi, R., Costa, G., Quaghebeur, M., Snellings, R.: Carbonate-bonded construction materials from alkaline residues. RILEM Tech. Lett. **2**, 53–58 (2017)
8. De Schutter, G., Taerwe, L.: General hydration model for Portland cement and blast furnace slag cement. Cem. Concr. Res. **25**(3), 593–604 (1995)
9. Nurse, R.W.: Steam curing of concrete. Mag. Concr. Res. **1**(2), 79–88 (1949)
10. Saul, A.G.A.: Principles underlying the steam curing of concrete at atmospheric pressure. Mag. Concr. Res. **2**(6), 127–140 (1951)
11. Higginson, E.C.: Effect of steam curing on the important properties of concrete. ACI J. Proc. **58**(9), 281–198 (1961)
12. De Schutter, G., Vuylsteke, M.: Minimisation of early age thermal cracking in a J-shaped non-reinforced massive concrete quay wall. Eng. Struct. **26**, 801–808 (2004)
13. Hedlund, H., Groth, P.: Air cooling of concrete by means of embedded cooling pipes-part I: laboratory tests of heat transfer coefficients. Mater. Struct. **31**(5), 329–334 (1998)
14. Groth, P., Hedlund, H.: Air cooling of concrete by means of embedded cooling pipes—part II: application in design. Mater. Struct. **31**(6), 387–392 (1998)
15. De Schutter, G.: Finite element simulation of thermal cracking in massive hardening concrete elements using degree of hydration based material laws. Comput. Struct. **80**(27–30), 2035–2042 (2002)
16. Rastrup, E.: Heat of hydration in concrete. Mag. Concr. Res. **6**(17), 79–92 (1954)
17. ACI Committee 308: Guide to external curing of concrete. ISBN 978-1-942727-87-3 (2016)
18. Kovler, K., Jensen, O.M. (eds.): Internal curing of concrete, state-of-the-art report of RILEM technical committee 196-ICC, RILEM Report 41. 978-2-35158-009-7 (2007)
19. Feys, D., Roussel, N., Verhoeven, R., De Schutter, G.: Influence of air content on the steady state rheological properties of fresh self-compacting concrete, without air entraining agents. In: Caijun Shi, Zhiwu Yu, Kamal Henri Khayat, Peiyu Yan (eds.) In: Proceedings of the 2nd International Symposium Design, Performance and Use of Self-Consolidating Concrete (SCC 2009). RILEM Publications, Beijing (China), vol 65, pp. 287–295, ISBN 978-2-35158-073-8 (5–7 June 2009)
20. Feys, D., De Schutter, G., Khayat, K., Verhoeven, R.: Changes in rheology of self-consolidating concrete induced by pumping. Mater. Struct. **49**, 4657–4677 (2016)
21. Dils, J., Boel, V., De Schutter, G.: Vacuum mixing technology to improve the mechanical properties of ultra-high performance concrete. Mater. Struct. **48**, 3485–3501 (2015)
22. Peters, S.: The influence of powder ultrasound on setting and strength development of cement suspensions. PhD thesis, Bauhaus University Weimar. ISBN 978-3-00-055602-9 (2017)
23. Latham, J.: Vacuum treatment of concrete. Civ. Eng., 15–21 (1979)
24. Yazıcı, H., Yardımcı, M.Y., Aydın, S., Karabulut, A.Ş.: Mechanical properties of reactive powder concrete containing mineral admixtures under different curing regimes. Constr. Build. Mater. **23**(3), 1223–1231 (2009)
25. Wang, Y., Wei, H., Li, Z.: Effect of magnetic field on the physical properties of water. Results Phys. **8**, 262–267 (2018)
26. Reddy, V.S., Kumar, A.K., Sumanth, A.: Effect of magnetic field treated water on fresh and hardened properties of concrete. J. Civ. Eng. Env. Techn. **4**(2), 134–138 (2017)
27. Su, N., Wu, C.-F.: Effect of magnetic field treated water on mortar and concrete containing fly ash. Cem. Concr. Compos. **25**(7), 681–688 (2003)

28. Soto-Bernal, J.J., Gonzales-Mota, R., Rosales-Candelas, I., Ortiz-Lozano, J.A.: Effects of static magnetic fields on the physical, mechanical, and microstructural properties of cement pastes. Adv. Mater. Sci. Eng., 934195, 1–9 (2015)
29. Nair, S.D., Ferron, R.D.: Set-on-demand concrete. Cem. Concr. Res. **57**, 13–27 (2014)
30. Abavisani, I., Rezaifar, O., Kheyroddin, A.: Alternating magnetic field effect on fine-aggregate steel chip-reinforced concrete properties. J. Mater. Civ. Eng. **30**(6), 04018087 (2018)
31. Rezaifar, O., Abavisiani, I., Kheyroddin, A.: Magneto-electric active control of scaled-down reinforced concrete columns. ACI Struct. J. **114**(5), 1351–1362 (2017)
32. Ponsot, I.M.M.M., Pontikes, Y., Baldi, G., Chinman, R.K., Detsch, R., Boccaccini, A.R., Bernardo, E.: Magnetic glass ceramics by sintering of borosilicate glass and inorganic waste. Materials. **7**, 5565–5580 (2014)
33. Joffroy, E., Koulialias, D., Yoon, S., Partl, M.N., Studart, A.R.: Iron oxide nanoparticles for magnetically-triggered healing of bituminous materials. Constr. Build. Mater. **112**, 497–505 (2016)
34. Joffroy, E., Bouville, F., Bueno, M., Studart, A.R., Partl, M.N.: Iron-based particles for the magnetically-triggered crack healing of bituminous materials. Constr. Build. Mater. **164**, 775–782 (2018)
35. Kovtun, M., Ziolkowski, M., Shekhovtsova, J., Kearsley, E.: Direct electric curing of alkali-activated fly ash concretes: a tool for wider utilization of fly ashes. J. Clean. Prod. **133**, 220–227 (2016)
36. Zhang, P., Chen, K., Wang, Y., Liu, Y.: Alkali-activated cementing material capable of controlling setting time as well as control method and application thereof. Patent CN105731947 A (6 July 2016)
37. Zhang, P., Chen, K., Wang, Y., Liu, Y.: Concrete construction technique capable of controlling setting time and special equipment therefor. Patent WO2016165413 A1 (20 Oct 2016)
38. Goudjil, N., Vanhove, Y., Djelal, C., Kada, H.: Electro-osmosis applied for formwork removal of concrete. J. Adv. Concr. Technol. **10**, 301–312 (2012)
39. Goudjil, N., Djelal, C., Vanhove, Y., Kada, H., Heloun, N.: Impact of temperature on the demoulding of concrete elements with a polarization process. Constr. Build. Mater. **54**, 402–412 (2014)
40. Choi, M.S., Kim, Y.S., Kim, J.H., Kim, J.S., Kwon, S.H.: Effects of an externally imposed electromagnetic field on the formation of a lubrication layer in concrete pumping. Constr. Build. Mater. **31**, 18–23 (2014)
41. Chess, P.M., Broomfield, J.P.: Cathodic Protection of Steel in Concrete, CRC Press. ISBN 9780203223031 (2003)
42. Pourbaix, M.: Atlas of Electrochemical Equilibrium in Aqueous Solutions. Pergamon, London (1976)
43. Xuequan, W., Jianbgo, D., Mingshu, T.: Microwave curing technique in concrete manufacture. Cem. Concr. Res. **17**, 205–210 (1987)
44. Leung, C.K.Y., Pheeraphan, T.: Microwave curing of Portland cement concrete: experimental results and feasibility for practical applications. Constr. Build. Mater. **9**(2), 67–73 (1995)
45. Steinschulte, A.A., Scotti, A., Rahimi, K., Nevskyi, O., Oppermann, A., Schneider, S., Bochenek, S., Schulte, M.F., Geisel, K., Jansen, F., Jung, A., Mallmann, S., Winter, R., Richtering, W., Wöll, D., Schweins, R., Warren, N.J., Plamper, F.A.: Stimulated transitions of directed nonequilibrium self-assemblies. Adv. Mater. **29**, 1703495 (2017)
46. Huo, M., Yuan, J., Tao, L., Wei, Y.: Redox-responsive polymers for drug delivery: from molecular design to applications. Polym. Chem. **5**, 1519–1528 (2014)
47. Zhu, L.Z., Shangguan, Y.G., Sun, Y.X., Li, J.A., Zhen, Q.A.: Rheological properties of redox-responsive, associative ferrocene-modified branched poly(ethylene imine) and its modulation by beta-cyclodextrin and hydrogen peroxide. Soft Matter. **6**(21), 5541–5546 (2010)
48. Jiang, J.Z., Zhang, D.Y., Yin, J.C., Cui, Z.G.: Responsive, switchable wormlike micelles for CO2/N-2 and redox dual stimuli based on selenium-containing surfactants. Soft Matter. **13**(37), 6458–6464 (2017)

49. Ramtin, A., Seyfoddin, A., Coutinho, F.P., Waterhouse, G.I.N., Rupenthal, I.D., Svirskis, D.: Cytotoxicity considerations and electrically tunable release of dexamethasone from polypyrrole for the treatment of back-of-the-eye conditions. Drug. Deliv. and Transl. Res. **6**, 793–799 (2016)
50. Wajs, E., Nielsen, T.T., Larsen, K.L., Fragoso, A.: Preparation of stimuli-responsive nano-sized capsules based on cyclodextrin polymers with redox or light switching properties. Nano Res. **9**(7), 2070–2078 (2016)
51. Sen, S., Palmore, G.T.R.: Stimuli-responsive macromolecular composites: enhanced stress modulation in polypyrrole with redox-active dopants. Macromolecules. **49**, 8479–8488 (2016)
52. Darabi, A., Jessop, P.G., Cunningham, M.F.: CO2-responsive polymeric materials: synthesis, self-assembly, and functional applications. Chem. Soc. Rev. **45**, 4391–4436
53. Oechsle, A.L., Lewis, L., Hamad, W.Y., Hatzikiriakos, S.G., MacLachlan, M.J.: CO2-switchable cellulose nanocrystal hydrogels. Chem. Mater. **30**, 376–385 (2018)
54. Manouras, T., Vamvakaki, M.: Field-responsive materials: photo-, electro-, magnetic-and ultrasound-sensitive polymers. Polym. Chem. **8**, 74–96 (2017)
55. Paris, J.L., Cabanas, M.V., Manzano, M., Valet-Regi, M.: Polymer-grafted mesoporous silica nanoparticles as ultrasound-responsive drug carriers, acsnano. Am. Chem. Soc., 11023–11033 (2015)
56. Du, B., Ren, X., Chang, R., Dong, Z.T.: An end-bifunctionalized hydrophobically modified ethoxylated urethane model polymer: multiple stimuli-responsive aggregation and rheology in aqueous solution. Macromolecules. **50**, 1688–1699 (2017)
57. Fang, H., Zhao, X., Lin, Y., Yang, S., Hu, J.: A natural glycyrrhizic acid-tailored light-responsive gelator. Chem. Asian J. **13**, 1192–1198 (2018)
58. He, L., Hu, B., Henn, D.M., Zhao, B.: Influence of cleavage of photosensitive group on thermally induced micellization and gelation of a doubly responsive diblock copolymer in aqueous solutions: a SANS study. Polymer. (2016). https://doi.org/10.1016/j.polymer.2016.10.019
59. Karimi, M., Ghasemi, A., Zangabad, P.S., Rahighi, R., Basri, S.M.M., Mirshekari, M., Amiri, M., Pishabad, Z.S., Aslani, A., Bozorgomid, M., Ghosh, D., Beyzavi, A., Vaseghi, A., Aref, A.R., Haghani, L., Bahrami, S., Hamblin, M.R.: Smart micro/nanoparticles in stimulus-responsive drug/gene delivery systems. Chem. Soc. Rev. **45**, 1457–1501 (2016)
60. Ma, X., Tian, H.: Stimuli-responsive supramolecular polymers in aqueous solution. Acc. Chem. Res. **47**, 1971–1981 (2014)
61. Roy, D., Cambre, J.N., Sumerlin, B.S.: Future perspectives and recent advances in stimuli-responsive materials. Prog. Polym. Sci. **35**, 278–301 (2010)
62. Brassinne, J., Fustin, C.A., Gohy, J.F.: Control over the assembly and rheology of supramolecular networks via multi-responsive double hydrophilic copolymers. Polym. Chem. **8**, 1527–1539 (2017)
63. Morales, D., Palleau, E., Dickey, M.D., Velev, O.D.: Electro-actuated hydrogel walkers with dual responsive legs. Soft Matter. **10**(9), 1337–1348 (2014)
64. Sankoh, S., Vagin, M.Y., Sekretaryova, A.N., Thavarungkul, P., Kanatharana, P., Mak, W.C.: Colloid electrochemistry of conducting polymer: towards potential-induced in-situ drug release. Electrochim. Acta. **228**, 407–412 (2017)
65. Okuzaki, H., Funasaka, K.: Electro-responsive polypyrrole film based on reversible sorption of water vapor. Synth. Metals. **108**, 127–131 (2000)
66. Huang, Y., Yong, P., Chen, Y., Gao, Y., Xu, W., Lv, Y., Yang, L., Reis, R.L., Pirraco, R.P., Chen, J.: Micellization and gelatinization in aqueous media of pH-and thermos-responsive amphiphilic ABC (PMMA82-b-PDMAEMA150-b-PNIPAM65) triblock copolylmer synthesized by consecutive RAFT polymerization. RSC Adv. **7**, 28711–28722 (2017)
67. van der Asdonk, P., Hendrikse, H.C., Sauli, A.C., Kraaijkamp, S.P.M., Kouwer, P.H.J.: Spatial and temporal patterning of polymers in electric field responsive LC templates. J. Mat. Chem. **4**(35), 8263–8269 (2016)
68. Cairns, D.R., Shafran, M.S., Sierros, K.A., Huebesch, W.W., Kessman, A.J.: Stimulus-responsive fluidic dispersions of rod shaped liquid crystal polymer colloids. Mater. Lett. **64**, 1133–1136 (2010)

69. Frechet, J.M.: Functional polymers and dendrimers: reactivity, molecular architecture, and interfacial energy. Science. **263**(5154), 1710–1715 (1994)
70. Jeon, S.J., Hauser, A.W., Hayward, R.C.: Shape-morphing materials from stimuli-responsive hydrogel hybrids. Acc. Chem. Res. **50**, 161–169 (2017)
71. Attaran, A., Brummund, J., Wallmersperger, T.: Modeling and simulation of the bending behavior of electrically-stimulated cantilevered hydrogels. Smart Mater. Struct. **24**, 035021 (2015)
72. Meng, H., Li, G.: A review of stimuli-responsive shape memory polymer composites. Polymer. **54**, 2199–2221 (2013)

Glossary

Geert De Schutter and Jay Sanjayan

Abstract This chapter presents an overview of relevant keywords that are further used in this state-of-the-art report, as well as in relevant literature on the topic of Active Control of Properties of Fresh and Hardening Cementitious Materials. For several of these keywords, the same definition is taken as already included in a previous RILEM State-of-the-Art Report [1]. Additional keywords have been included in the list as well, giving a definition formulated in consensus within the RILEM Committee responsible for the development of current state-of-the-art report.

Keywords Active control · Cementitious materials · Fresh properties · Hardening properties

Active Control

Active control of properties of cementitious materials refers to the action of controlling some specific properties (e.g. rheology, stiffening, ...) by applying an external trigger signal that induces some response in the cementitious material, with a modification of properties or behaviour as a result. In the ideal active control process, the intervention is based on observations and/or measurements (quantified), giving the necessary information to take a decision regarding required intervention. A feed-back loop between measurement and signal seems an optimal situation.

G. De Schutter (✉)
Magnel-Vandepitte Laboratory, Department of Structural Engineering and Building Materials, Ghent University, Ghent, Belgium
e-mail: Geert.DeSchutter@ugent.be

J. Sanjayan
Swinburne University of Technology, Melbourne, Australia

G. De Schutter and J. Sanjayan (eds.), *Active Control of Properties of Fresh and Hardening Cementitious Materials*, RILEM State-of-the-Art Reports 50,
https://doi.org/10.1007/978-3-032-30340-0_2

Addition

A fine-grained inorganic material. Includes two types [EN206]: inert or nearly inert addition (Type I) and pozzolanic or latent hydraulic addition (Type II). (See also supplementary cementitious materials, fines, and fine fillers).

Admixture

A material, other than water, aggregate, hydraulic cement, and fibre reinforcement, used as an ingredient of a cementitious mixture to modify its freshly mixed, setting, or hardened properties and that is added to the batch before or during its mixing.

Air content

Volume of air-voids in fresh or hardened concrete usually expressed as a percentage of the total volume of the mix. Vibrated concrete is usually considered to be fully compacted when its air content is below 1.5%, excluding any entrained air.

Air-entrainment

Intentional introduction of air uniformly distributed in small air bubbles into concrete, primarily to improve frost resistance. Volume of entrained air in SCC usually varies between 5%—8% of the total volume of concrete.

Autogenous shrinkage

The external-macroscopical (bulk) dimensional reduction (volumetric or linear) of the cementitious system, which occurs under sealed, isothermal, and unrestrained conditions.

Binder

A cementitious material, either a hydrated cement or reaction product of cement or lime and reactive siliceous material; the kind of cement and curing conditions governing the characteristics of the product formed. Also materials such as asphalt, resins, and other materials forming the matrix of concrete, mortar, and sanded grout.

Bingham fluid

A material which when subjected to shear stress behaves as an elastic solid until the yield stress is reached, after which there is a linear variation between shear stress and rate of strain (velocity gradient). The slope of the linear relationship refers to the plastic viscosity. Fresh cement-based materials at relatively low shear rate values and early age typically behave as Bingham fluids.

Bleed water

The water that rises to the surface subsequent to the placing of the concrete. The rise of mixing water within, or its emergence from, newly placed concrete, can be caused by consolidation and settlement of the plastic concrete.

Bleeding test

The standard test for determining the relative quantity of mixing water that will bleed from a sample of freshly mixed concrete.

Blocking

The condition in which coarse aggregate particles combine to form elements large enough to obstruct the flow of the fresh concrete among closely spaced reinforcement and/or other obstructions in the formwork.

Cohesiveness

The tendency of the concrete constituent materials to adhere together, resulting in resistance to segregation, settlement, and bleeding.

Compaction

The process in which the volume of entrapped air in fresh concrete is reduced below 1%–2% of total volume of concrete, usually by mechanical means, such as vibration. Interchangeable with "consolidation". In SCC, compaction (consolidation) is achieved by gravity flow of the material without the need of mechanical vibration, rodding, or tamping.

Compressive strength

The capacity to withstand axially directed pushing stresses. When the limit of compressive strength is reached, the concrete is crushed.

Consolidation

See compaction.

Conventional vibrated concrete

See vibrated concrete.

Consistency

The relative mobility or ability of freshly mixed concrete or mortar to flow.

Drying shrinkage

Shrinkage occurring in a specimen that is exposed to the environment and allowed to dry.

Fiber

A slender or threadlike component sometimes used to strengthen cementitious materials, e.g. metal, synthetic, or natural.

Fillers

Finely divided inert materials, such as limestone powder, silica, or colloidal substances, sometimes added to Portland cement, paint or other materials to reduce shrinkage, improve workability, or act as an extender or material used to fill an opening in a form.

Filling ability

The ability of SCC to flow into the formwork and fill completely all spaces within the formwork under its own weight, also referred to as deformability or non-restricted deformability. It is sometimes referred to as 'flowability'.

Fines and fine fillers

Includes powder (see Powder) and materials finer than 0.125 mm (No. 100 sieve).

Flexural strength

Highest tensile stress experienced within the material at its moment of rupture in flexure.

Flowability

See Filling ability.

Flow-rate

The speed at which a sample of a fresh concrete mixture spreads horizontally outwards from the base of a slump cone until it reaches a selected radial distance, usually a concentric ring of 500 mm diameter. It is indicated by a 'flow-time', such as t_{50}, interchangeable with t_{500}, measured in second.

Flow-time

Usually, the time taken for the concrete flow in an Orimet, V-funnel, or O-funnel test; an indication of the level of filling ability and rate of flow. Flow-time can be used to assess plastic viscosity and risk of blocking of SCC.

Fluidity

The ease by which fresh concrete flows under gravity. In rheology, fluidity is the reciprocal value of viscosity.

Fly ash

The finely divided residue that results from the combustion of ground or powdered coal and that is transported by flue gasses from the combustion zone to the particle removal system. Because of its spherical shape, fineness, and pozzolanic activity, fly ash can modify the fresh and hardened properties of concrete.

Formwork leakage

The loss of cement paste or fine mortar fraction through defects or joints in formwork.

Formwork pressure

Lateral pressure acting on vertical or inclined formed surfaces, resulting from the fluid-like behaviour of the unhardened concrete confined by the forms.

Ground granulated blast-furnace slag (GGBFS, also known as GGBS)

A fine granular, mostly latent hydraulic binding material that can be added to concrete to modify its fresh and hardened characteristics. A waste product in the manufacture of pig iron and chemically a mixture of lime, silica, and alumina. GGBFS is also referred to in some cases as slag cement.

High-range water-reducing admixture (HRWRA)

A water-reducing admixture capable of producing large water reduction or greater flowability, also known as superplasticiser.

Metakaolin

Mineral admixture used as binding material (supplementary cementitious material) in concrete. A highly processed reactive aluminosilicate pozzolan, a finely-divided material that reacts with slaked lime at ordinary temperature and in the presence of moisture to form strong slow-hardening cement. It is formed by calcining purified kaolinite, generally between 650–700 °C in an externally fired rotary kiln.

Mixture robustness

The characteristic of a mixture that encompasses its tolerance to variations in constituent characteristics and quantities, variations during concrete mixing, transport, and placement, as well as environmental conditions.

Modulus of elasticity

The slope of the stress-strain curve in the elastic deformation region.

Mortar

The fraction of concrete with maximum aggregate size ≤4 mm. The maximum size can be different (e.g. 4.75 mm as per ASTM, ACI, and CSA specifications).

Newtonian fluid

A fluid, which shows a linear relationship between shear rate and shear stress, passing through the origin in a simple shear flow. Fresh SCC mixtures of high filling ability can display this type of behaviour. The slope of the linear relationship is the Newtonian viscosity (or viscosity).

Paste

Mixture of cementitious materials and fillers with water, with or without chemical admixtures.

Paste volume

Volume of paste in concrete or in mortar; it can be expressed as volume percent of the entire mixture.

Plastic settlement

Settlement of a concrete after casting while the material is still in fresh state.

Plastic viscosity

The resistance of the plastic material to undergo a given flow. It is computed as the slope of the shear stress versus shear rate curve measurements. Concrete with higher plastic viscosity takes longer to flow. It is closely related to t_{50} and V-funnel time (higher plastic viscosity: higher t_{50} and V-funnel time). (See also Bingham fluid).

Powder

Includes cement, fillers, and additions.

Pumpability

The ability of a SCC mixture to be pumped without significant degradation of workability or blockage.

Rate of flow

See flow-rate.

Rheology

The science of flow and deformation of matter. It includes studies of the handling and placing of freshly mixed concrete, and the behaviour of slurries, pastes, and the like. In the context of SCC, rheology refers to the evaluation of yield stress, plastic viscosity, and thixotropy to achieve the desired level of filling ability, passing ability, and segregation resistance. Rheology is also used to evaluate deformation of hardened concrete.

Rheological properties

Properties dealing with the deformation and flow of matter.

Robustness

See mixture robustness.

Segregation

The differential concentration of the components of mixed concrete, aggregate, or the like, thus resulting in non-uniform proportions in the mass. In the case of SCC, segregation may occur during transport, flow into the form, or after placement when the concrete is in a plastic state. This results in non-uniform distribution of in-situ properties of the concrete.

Segregation resistance

The ability of concrete to remain uniform in terms of composition during placement and until setting. Segregation resistance encompasses both dynamic and static stability.

Self-compacting concrete (SCC)

Fresh concrete which has an ability to flow under its own weight, fill the required space or formwork completely and produce a dense and adequately homogeneous material without a need for mechanical compaction.

Self-consolidating concrete (SCC)

See self-compacting concrete.

Self-levelling concrete

SCC with the ability to attain a fully levelled surface after casting.

Service life

The time during which the structure performs its design functions with and without unforeseen maintenance or repair.

Settlement

The condition in which the aggregates in SCC tend to sink to the bottom of the form resulting in non-homogeneous concrete. Surface settlement can also be caused by bleeding of free water and loss of air as well as movement of aggregate particles within fresh concrete (consolidation).

Shear strain

A deformation of a material in which parallel internal surfaces slide past one another.

Shear thinning (also known as pseudo-plasticity)

A material is shear thinning when the viscosity is decreasing at increasing shear rate.

Shear thickening

A material is shear thickening when the viscosity is increasing at increasing shear rate.

Shear strength

The strength of a material when loaded in shear.

Shear stress

The stress component acting tangentially to a plane.

Shrinkage

Contraction of concrete due to loss of water, e.g. due to the hydration process (autogenous shrinkage) or due to water loss to the environment (drying shrinkage).

Shrinkage-reducing agent

Admixture added to cementitious materials to reduce the shrinkage.

Silica fume

Very fine amorphous silica produced in electric arc furnaces as a by-product of the production of elemental silicon or alloys containing silicon. Silica fume can be added to SCC to improve the rheological and mechanical properties.

Slump flow retention

The ability of concrete to maintain its slump flow over a given period of time.

Slump flow spread

The distance of lateral flow of concrete during the slump-flow test. Slump flow spread is the numerical value of flow determined as the average diameter of the circular deposit of SCC at the conclusion of the slump flow test.

Slump-flow test

A test for filling ability of SCC using a 300 mm tall conical mould (as for the slump test). The primary result is the average horizontal spread diameter (SF).

Slump test

A test for consistency of fresh traditional vibrated concrete. Result is expressed as the depth of the vertical settlement of concrete after removal of a standard (Abrams) mould 300 mm high.

Speed of flow

See flow-rate.

Spread

The result of the 'Slump-flow' (SF) or J-ring (SFJ) tests.

Stability

The ability of a concrete mixture to resist segregation of the paste from the aggregate.

Stability, Dynamic—The resistance to segregation when external energy is applied to concrete—namely during placement.

Stability, Static—The resistance to segregation when no external energy is applied to concrete—namely from immediately after placement and until the onset of hardening.

Stabilizer

Admixture applied to improve the stability of the mixture (See al-so viscosity-modifying admixture).

Strain

Geometrical measure of deformation representing the relative displacement between particles in a material.

Stress

Amount of force exerted per unit area.

Superplasticiser

An admixture producing much-increased consistency of a fresh mixture without significant retardation or air-entrainment. Also known as 'high-range water reducing admixture—HRWRA'.

Supplementary cementitious materials

See Additions; [EN206]: Type II pozzolanic or latent hydraulic addition.

Tensile strength

Stress at which a material breaks when loaded in tension.

Thixotropy

The property of a material that enables it to stiffen in a short period while at rest (structural build-up), but to acquire a lower viscosity when mechanically agitated (structural break-down); the process being time-dependent and reversible. The material having this property is termed thixotropic.

Traditional vibrated concrete

See vibrated concrete.

Transportability

The ability of concrete to be transported from the mixer to the placement site while remaining in a homogeneous condition.

Vibrated concrete

Concrete, which requires vibration to achieve adequate compaction or consolidation.

Viscosity (plastic)

See plastic viscosity.

Viscosity-modifying admixture (VMA)

An admixture used for enhancing the rheological properties of cement-based materials in the plastic state to reduce the risk of segregation and washout. Also known as thickening agent, stabilizer, or viscosity-enhancing agent (VEA).

Water-cement ratio (W/C)

The ratio of the mass of water, exclusive only of that absorbed by the aggregate, to the mass of cement in concrete, mortar, or grout.

Water-cementitious materials ratio (W/CM)

The ratio of the mass of water, exclusive only of that absorbed by the aggregate, to the mass of cementitious materials (cement and addition Type II) in concrete, mortar, or grout. Also referred to as water-binder ratio (W/B).

Water-powder ratio (W/P)

The ratio of the mass of water, exclusive only of that absorbed by the aggregate, to the mass of cement and other powder additions (Types I and II) in concrete, mortar, or grout.

Workability

That property of freshly mixed concrete or mortar that determines the ease with which it can be mixed, placed, consolidated, and finished into a homogenous condition. For SCC, workability encompasses filling ability, passing ability, and segregation resistance, and it is affected by rheology. (See also consistency).

Yield stress

The minimum shear stress required to initiate (static yield stress) or maintain (dynamic yield stress) flow. The dynamic yield stress is closely related to slump flow of SCC (lower yield stress results in higher slump flow). Dynamic yield stress is calculated as the intercept of the shear stress versus shear rate plot from rheometer flow curve measurements. Also known as yield value, the stress on the stress-strain or shear stress—rate of shear strain diagram which corresponds to the change from elastic to plastic behaviour of a solid material or static to dynamic behaviour of a Bingham fluid.

Reference

1. Khayat, K.H., De Schutter, G.: Mechanical properties of self-compacting concrete. In: RILEM State-of-the-Art Reports, vol. 14. Springer (2014). https://doi.org/10.1007/978-3-319-03245-0

Active Control of Rheology of Cementitious Materials

Dengwu Jiao, Geert De Schutter, Ángel de la Rosa Velasco, Anil Kumar Kannauzia, Luise Göbel, Onur Ozturk, Qiang Yuan, Paula Heik, Shravan Muthukrishnan, Sriramya Nair, Swapnil Balasaheb Ghodke, Thomas Kränkel, Ye Qian, Yi Zhang, Yiyuan Zhang, and Noor Yaseen

Abstract Rheology plays an important role in influencing workability, placing process, microstructure, mechanical features, and durability of cementitious materials. Traditional rheology control approaches rely on adjusting mixture compositions through the incorporation of supplementary cementitious materials and the

D. Jiao (✉)
Department of Architecture and Civil Engineering, City University of Hong Kong, Kowloon Tong, Hong Kong SAR, China
e-mail: dengwu.jiao@cityu.edu.hk

G. De Schutter · Y. Zhang · Y. Zhang
Magnel-Vandepitte Laboratory, Department of Structural Engineering and Building Materials, Ghent University, Ghent, Belgium
e-mail: Geert.DeSchutter@ugent.be

Y. Zhang
e-mail: Yi.Zhang@ugent.be

Y. Zhang
e-mail: Yiyuan.Zhang@ugent.be

Á. de la Rosa Velasco
DIMME, Grupo de Durabilidad e Integridad Mecánica de Materiales Estructurales, Universidad Rey Juan Carlos, Móstoles, Madrid, Spain
e-mail: angel.delarosa@urjc.es

A. K. Kannauzia
Department of Civil Engineering, Sardar Vallabhbhai National Institute of Technology, Surat, Gujarat, India
e-mail: d20ce035@ced.svnit.ac.in

L. Göbel · P. Heik
Chair of Mechanics of Engineering Materials, Bauhaus-Universität Weimar, Weimar, Germany
e-mail: luise.goebel@uni-weimar.de

P. Heik
e-mail: paula.heik@uni-weimar.de

O. Ozturk · S. Nair
School of Civil and Environmental Engineering, Cornell University, Ithaca, NY, USA

G. De Schutter and J. Sanjayan (eds.), *Active Control of Properties of Fresh and Hardening Cementitious Materials*, RILEM State-of-the-Art Reports 50,
https://doi.org/10.1007/978-3-032-30340-0_3

addition of chemical admixtures. However, once the mixture proportion is determined and the concrete is prepared, no further interventions can be made to the rheological properties during the placement process. This chapter summarizes the recent advances on the active control of rheological properties of fresh cementitious materials using external trigger signals. The basics of concrete rheology, including viscoelasticity, yield stress, viscosity, thixotropy, shear thinning, and shear thickening were first described. Subsequently, various active rheology control methods induced by magnetic field, electric field, vibration, microwave, temperature, ultrasonic, pH/chemical, and pressure, are discussed in detail from the viewpoints of principles, influencing factors, and potential applications. At last, future research prospects in this new topic, as well as a summary, are provided.

Keywords Rheology · Thixotropy · Structural build-up · Magnetic Field · Temperature

1 Introduction

The engineering application of concrete includes critical processes of mixing, transporting, pumping and formwork casting. Each process is a significant factor influencing the properties of both fresh and hardened concrete [1, 2]. Take the pumping

e-mail: oo59@cornell.edu

S. Nair
e-mail: sn599@cornell.edu

Q. Yuan
School of Civil Engineering, Central South University, Changsha, China
e-mail: yuanqiang@csu.edu.cn

S. Muthukrishnan
TU Dresden, Dresden, Germany

S. B. Ghodke
Department of Civil Engineering, Indian Institute of Technology Roorkee, Roorkee, Uttarakhand, India
e-mail: swapnilb_ghodke@ce.iitr.ac.in

T. Kränkel
Department of Materials Engineering, Technical University of Munich, Munich, Germany
e-mail: thomas.kraenkel@tum.de

Y. Qian
Department of Civil Engineering, The University of Hong Kong, Hong Kong, China
e-mail: yq2157@columbia.edu

N. Yaseen
Department of Civil Engineering, University of Engineering and Technology, Lahore, Pakistan

Irrigation Department, Khyber Pakhtunkhwa, Pakistan

operation as an example, it has become an indispensable part of modern concrete construction practices. Over the years, extensive research has been conducted on the characterization of flow regimes within pipelines and the prediction of pumping pressure [3–5]. Despite these advancements, several challenges persist in the pumping operations, such as sensitivity to human factors, impacts of pipe bends, reducers, and temperature, aggregate segregation, and blockages [6, 7]. In addition, the pumping process inherently alters multiple properties of fresh concrete, such as yield stress, viscosity, air content and even temperature. These changes are often inevitable and hardly controllable, as the properties of fresh concrete are determined by its initial mixture proportions and mixing process, which cannot be modified in real time during the pumping process. Furthermore, conflicting requirements for fresh concrete performances arise across different operational processes. For example, low thixotropic structural build-up is essential for overcoming the challenges of resuming pumping operations after a short interruption (e.g. due to a delay by concrete trucks) [8] and improving the interface properties during multi-casting process [9]. In contrast, a higher structuration rate of concrete is beneficial for reducing formwork pressure during casting [10, 11]. The competing demands highlight the limitations of conventional static mixture designs and the potential of rheology control strategies that can adapt to process-specific needs.

Traditional approaches to rheology control primarily rely on adjusting mixture compositions through the incorporation of supplementary cementitious materials (SCMs) [12–14], such as fly ash, granulated blast furnace slag, and silica fume, and the addition of chemical admixtures (e.g., superplasticizers, air entraining agent, and viscosity modifying agent) [15–17]. While effective for initial tuning of rheological properties, these methods lack the ability to adjust fresh concrete properties in real time. That is, once the mixture proportion is fixed and the concrete is prepared, no further interventions can be made to the rheological properties during the placement process. To overcome these challenges, the concept of active rheology control has been introduced [18, 19], representing a fundamental shift from passive mixture design to dynamic real-time rheological regulations. Recent attempts have explored various active rheology control techniques that utilize external trigger signals such as magnetic field, vibration, temperature, and ultrasonic wave, which has been briefly summarized in a RILEM Technical Letter '*Active Control of Properties of Fresh and Hardening Concrete*' [20]. By enabling real-time adjustments of key rheological parameters, these methods offer solutions to long-standing challenges in pumping, casting, and emerging applications like 3D printing.

This chapter summarizes the recent advances on the active control of rheological properties of fresh cementitious materials. It begins by introducing the basics of concrete rheology, including viscoelasticity, yield stress, viscosity, thixotropy, shear thinning, and shear thickening. Subsequently, various active rheology control methods induced by magnetic field, electric field, vibration, microwave, temperature, ultrasonic, pH/chemical, and pressure, are discussed in detail from the viewpoints of principles, influencing factors, and potential applications. Finally, future research prospects in this new topic, as well as a summary, are given. This chapter hopefully gives a fundamental understanding and insightful inspiration into the active rheology control of fresh cement-based materials for researchers and scholars.

2 Basics About Rheology

Rheology, the study of the flow and deformation of materials, is crucial for understanding cementitious materials. The term of "rheology" was coined in the 1920s and was inspired by Greek words "rheo" and "logia", reflecting its focus on how materials change shape and flow under force. Cementitious materials like concrete and mortar have complex rheological properties affected by factors like cement paste composition, aggregates, admixtures and mixing/application processes [18]. Understanding the rheological properties of cement pastes, mortars, and concretes is important because these properties influence microstructure, mechanical features, durability, consistency, workability, and ease of casting or placement [21, 22]. The following sub-sections outline the fundamentals of concrete rheology, encompassing essential concepts including viscosity, yield stress, shear thinning and shear thickening, thixotropy, and viscoelasticity. By exploring these key rheological principles, we can gain a more comprehensive understanding of the critical role that rheology plays in governing the behaviour and performance of cementitious materials in a range of construction applications.

2.1 Viscoelasticity

Viscoelasticity is a property of materials that exhibit both viscous and elastic behaviour when undergoing deformation. The viscoelastic behaviour of cementitious materials is crucial for predicting and controlling their fresh- and hardened-state performances in various construction applications, including transport, pumping, casting, and placement. Their behaviour depends on the curing time and material composition [23]. At short times or high loading frequencies, they act like elastic solids, storing energy and showing reversible deformation. At longer times or lower frequencies, they act more like viscous fluids, dissipating energy through irreversible flow.

The viscoelastic nature of cementitious materials is a result of their complex microstructure, which exhibits both viscous and elastic characteristics. When subjected to an applied stress or strain, these materials display a combination of viscous flow and elastic deformation, leading to a time-dependent response that is characteristic of viscoelastic behaviour.

Cement paste is a viscoelastic material that can be probed in its elastic regime without permanent destruction. To analyse the viscoelastic behaviour of cement paste, a strain oscillation can be applied, and the resulting stress response can be measured. The strain oscillation, as analysed through strain-controlled amplitude test, can be represented by the equation:

$$\gamma(t) = \gamma_0 \cdot \sin(\omega t) \tag{1}$$

where $\gamma(t)$ is the time-dependent strain, γ_0 is the strain amplitude, ω is the angular frequency in rad/s, and t is time in s. The resulting stress response can be represented by the equation:

$$\tau(t) = \tau_0 \cdot \left(G' \cdot \sin(\omega t) + G'' \cdot \cos(\omega t)\right) \tag{2}$$

where $\tau(t)$ is the time-dependent stress, G′ is the elastic modulus (or storage modulus) in Pa, and G″ is the viscous modulus (or loss modulus) in Pa. The elastic modulus G′ represents the elastic effects that are in phase with the strain, while the viscous modulus G" represents the viscous effects that occur in phase with the strain rate. The elastic (G′) and viscous (G″) modulus collectively constitute the complex modulus (G*) in Pa.

$$G^* = G' + iG'' \tag{3}$$

where i is the imaginary unit. The phase angle δ can be calculated as:

$$\tan\delta = G''/G' \tag{4}$$

In an ideal solid, where elasticity is perfect, the loss modulus (G″) is zero, and the complex modulus (G*) is equal to the storage modulus (G′). Conversely, in an ideal liquid, where rigidity is absent, the storage modulus (G′) is zero, and the complex modulus (G*) is equal to the loss modulus (G″). For cementitious pastes, at low strains, the storage modulus dominates, while at higher strains, the viscous modulus becomes more prominent, showcasing the material's complex viscoelastic response. Simultaneously measuring the elastic modulus and viscous modulus is essential for comprehensively characterizing the viscoelastic behaviour of cementitious materials [19].

The viscoelastic behaviour of cementitious materials is influenced by various factors, including the use of chemical admixtures [24]. These admixtures can impact the interparticle interactions and the stability of the microstructural network, thereby affecting the reversible breakdown and reformation processes that drive the thixotropic response. The ongoing hydration process also influences the viscoelastic behaviour by altering the material's microstructure.

2.2 *Yield Stress and Plastic Viscosity*

Yield stress refers to the minimum shear stress required for a non-Newtonian fluid to begin to change from a solid-like state (or static state) to a flowing state. For a fluid with yield stress, when the applied shear stress is less than its yield stress, the fluid will behave as a solid or nearly static state; when the applied shear stress exceeds the yield stress, the fluid will begin to flow. The inherent material resistance shown in the rheological models as the intercept with the vertical axis is called yield stress [1], which is said to originate from the spatial network of cement particles and the behaves elastically under applied stresses below the yield stress [25]. The Bingham model and Herschel-Bulkley model are commonly applied to describe the rheological behaviour of cement-based materials [26].

The plastic viscosity of cementitious materials is an important parameter to measure its flow resistance, which indicates the internal friction or fluid resistance of the materials when it flows at a certain shear rate. It reflects the difficulty of cementitious materials to maintain flow after being subjected to external forces. Cementitious materials normally (at least within the range of lower shear rates) can be described by the Bingham model simply, which is as follows:

$$\tau = \tau_0 + \mu\dot{\gamma} \tag{5}$$

where τ is shear stress, τ_0 is yield stress, μ is plastic viscosity, and $\dot{\gamma}$ is shear rate. Several factors such as water-to-cement ratio, admixtures, particle/aggregate properties, and external properties influence the yield stress and plastic viscosity of fresh cementitious materials. Note that the rheological properties are also influenced by the characteristics of the rheometer, testing protocol, temperature, pressure, etc.

2.3 *Thixotropy*

The thixotropy refers to the (reversible) change of the rheological properties of cementitious materials such as viscosity and yield stress with flow history [27, 28]. A decrease in viscosity under shear, followed by an increase upon removal of shear, is commonly observed. Roussel et al. [25] assumed that hydration products, even at early ages, contribute to the structuration of fresh cement paste. Early hydration products on the surface of cement particles bridge these particles to form a rigid network, which exhibits a relatively small critical strain. By contrast, colloidal forces of cement powders contribute to a softer network with a relatively higher critical strain. Thus, thixotropy is closely related to flocculation state of cement particles and/or agglomerates.

For plain cement pastes, thixotropy could be described as paste deflocculation and breaking down of C-S-H bridges under flow, and flocculation and formation of C-S-H bridges over time at rest [29, 30]. It has been observed that under constant shear rate, the gradual stress decay is related to the deflocculation process of cement pastes where bigger agglomerates are broken down to smaller particles/agglomerates [31]. Static yield stress (the minimum shear stress to initiate the flow) and dynamic yield stress (the minimum shear stress to sustain the flow) are found to correspond to the microstructural state of fresh materials before and after structural breakdown. It has been reasoned that thixotropy is related to the discrepancy between static yield stress and dynamic yield stress [32]. Some applications such as 3D concrete printing require a good balance in pumpability and buildability. Therefore, high thixotropy is desired with low dynamic yield stress under flow while maintaining high static yield stress under rest.

2.4 *Shear Thinning and Shear Thickening*

Shear thickening and shear thinning are two common rheological behaviours observed in non-Newtonian fluids [33]. Under shear, certain fluids exhibit a decrease in viscosity with increasing shear rate, known as shear thinning. Conversely, other fluids demonstrate an increase in viscosity as the shear rate increases, referred to as shear thickening. Interestingly, some specially formulated systems or those containing specific admixtures may display shear thickening at low shear rates and transition to shear thinning at higher shear rates, demonstrating a "reversed" shear response. These rheological phenomena are commonly observed in both natural and industrial systems, including starch suspensions, concentrated colloids, blood, debris flows, and polymer solutions [34]. Generally, shear thinning is attributed to the alignment of particles along the direction of shear or the breakdown of internal structures, which reduces internal friction, as shown in Fig. 1. In contrast, shear thickening is often associated with short-range particle aggregation, the formation of transient networks, the breakage of functional groups or side chains of the superplasticizer polymer, or intensified hydrodynamic interactions under high shear conditions [35, 36].

Traditional Bingham model is no longer sufficient to accurately describe the complex nonlinear characteristics associated with shear thinning and shear thickening phenomena. As a result, the Herschel–Bulkley model is widely used to describe

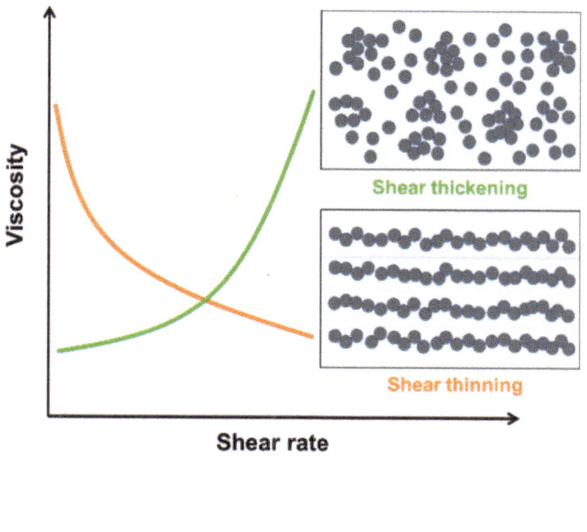

Fig. 1 Viscosity in shear thickening and shear thinning fluids (Courtesy of Dengwu Jiao)

this nonlinear relationship. Its expression is given as follows:

$$\tau = \tau_y + k\dot{\gamma}^n \tag{6}$$

where K is the consistency index, Pa s^n, n is the flow index. When $n < 1$, the system exhibits shear thinning behaviour; when $n > 1$, it exhibits shear thickening behaviour. If $n = 1$, the model reduces to the classical Bingham model. Besides, the Herschel–Bulkley model does not directly yield the plastic viscosity, but an equivalent plastic viscosity (μ_{eq}) can be calculated using the consistency index, the flow behaviour index, and the maximum shear rate ($\dot{\gamma}_{max}$), as expressed below:

$$\mu_{eq} = \frac{3k}{n+2}\dot{\gamma}_{max}^{n-1} \tag{7}$$

As a typical suspension system, fresh cement-based materials also exhibit pronounced shear thinning or shear thickening behaviour under certain conditions. Most cement slurries display shear thinning characteristics, where particles remain in a relatively loose suspended state and the viscosity decreases with increasing shear rate. However, under specific mix proportions or with the addition of chemical admixtures, shear thickening may occur at high shear rates. For example, Huang et al. [37] observed that in self-compacting concrete (SCC), the incorporation of superplasticizers (SP) led to shear thickening behaviour within a certain range of shear rates. They attributed this behaviour to two primary mechanisms: (1) desorption of SP under shear stress, which promotes the aggregation of cement particles; and (2) increased disorder of polymer chains and particles under high shear. Further research by Chia and Zhang [38] showed that the use of air-entraining agents (AEA) also significantly influences the shear response of cementitious slurries. When the air content of the mixture exceeds a certain threshold, the sample transitions from shear thickening to shear thinning. This indicates that the presence of entrained air bubbles with lubrication effect can inhibit particle clustering.

The shear thinning and shear thickening behaviours of cement-based materials are influenced by various factors, including the water-to-binder (w/b) ratio, temperature, mixing method, type of admixtures, and raw material properties [39–43]. A low w/b ratio results in a higher particle concentration and stronger interparticle interactions, which promotes the formation of flocculated structure and may induce shear thickening. In contrast, a higher w/b ratio enhances dispersion within the system, making shear thinning more likely to occur. Elevated temperatures accelerate the hydration reaction and may lead to the early formation of structural networks, potentially triggering shear thickening under moderate to high shear rates. Conversely, at lower temperatures, the slower reaction rate favours increased flowability. Inadequate mixing can cause local particle accumulation and localized thickening, while excessive mixing may disrupt the flocculated structure, facilitating shear thinning. What's more, the type and dosage of chemical admixtures, as well as the physical characteristics of raw materials, such as particle size, shape, and density, can significantly affect the shear response of the slurry [44].

Therefore, a thorough understanding and effective control of the shear thinning and shear thickening behaviour of cement-based materials are of great importance for engineering applications as well as the processing and shaping of these materials. For instance, during pumping, shear thinning behaviour can significantly reduce viscosity under applied shear stress, thereby lowering pipeline resistance, decreasing pumping pressure, and improving construction efficiency. Meanwhile, in processes such as formwork filling, self-supporting structural formation, or 3D printing, moderate shear thickening can help the material rapidly regain high viscosity and static stability once the shear ceases, thus preventing collapse and maintaining shape fidelity.

3 Active Rheology Control Techniques

3.1 *Magneto-Induced Rheology Control*

The possibility to control rheology of cementitious materials by applying a magnetic field is based on the fundamental principles of magnetorheological (MR) fluids. MR fluids, first introduced by Rabinow in 1948 [45], are a class of smart materials consisting basically of dispersed magnetic particles within a carrier fluid. In the absence of a magnetic field, these magnetic particles are randomly distributed, allowing the fluid to exhibit Newtonian behaviour. Upon applying an external magnetic field, however, the dipoles inside the magnetic particles become magnetized, causing them to align along the direction of the magnetic field and form chain-like or columnar structures [46–48]. As a consequence, the MR fluid transforms from liquid to semi-solid state within milliseconds, where the viscosity could increase magnitudes of 10^5–10^6 times of its original value and the field-induced yield stress can rise up to 100 kPa [49]. Once the external magnetic field is removed, the magnetic particles are demagnetized and the MR fluid reversibly transitions back to its liquid state, as shown in the schematic diagram in Fig. 2. Therefore, due to their unique ability to switch between fluid-like and solid-like states, MR fluids have found extensive applications in various engineering domains, including dampers, brakes, clutches, polishing, valves, and even medical devices [50, 51]. Despite their versatility, a major limitation of conventional MR fluids is the sedimentation of magnetic particles, which occurs due to their big density difference with the carrier fluid. This sedimentation poses a challenge for industrial applications, necessitating the development of stabilization techniques such as particle surface coatings, addition of chemical additives, and use of yield-stress fluids as carrier media. Fresh cementitious paste is usually regarded as a concentrated suspension exhibiting yield stress behaviour [52, 53]. When employed as the carrier fluid for MR systems, it offers improved stability compared to traditional MR fluids, mitigating sedimentation issues. Conversely, the rheological properties and stiffness of fresh cement paste containing magnetic particles would become controllable when subjected to electromagnetic interventions [54].

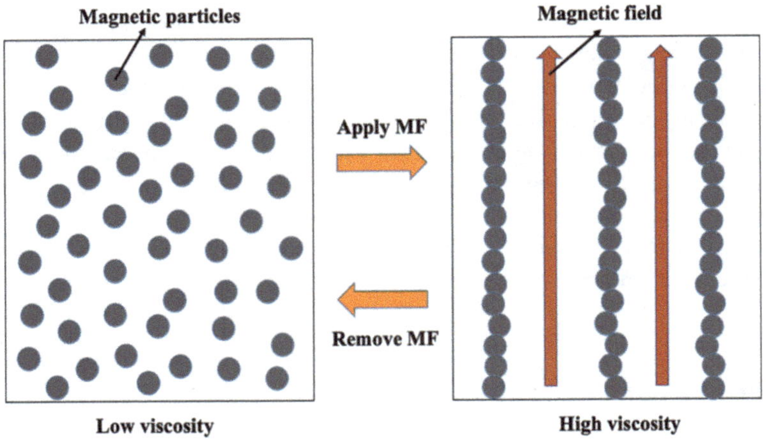

Fig. 2 Diagram illustrating the reversibility of MR fluids (Courtesy of Dengwu Jiao)

High magnetic saturation and low remanence magnetic powders are mainly used in magnetorheological fluids, enabling effective and reversible rheological control in real time. Among commonly used materials, different grades of microscale carbonyl iron powder (CIP) and nanoscale iron oxides are popular for providing magnetorheological properties to cementitious materials. These powders vary in particle size distributions and magnetic responses. Researchers have explored how different magnetic particles impact the stiffness control of cementitious materials. Nair and Ferron [55] compared the use of two grades of CIP in a cementitious mixture, with the value of D_{50} (the particle size corresponding to 50% of the particles being smaller than the given value) of 3.3 and 8.7 μm, and found that they exhibit similar yield stress, storage, and loss modulus across magnetic flux densities ranging from 0.2 to 1.0 T.

Jiao et al. [56, 57] reported stronger magnetorheological effects in mixtures containing larger and more responsive particles (i.e., nano-Fe_3O_4 particles). Under the application of an external magnetic field, the magnetic particles in cementitious suspensions will try to move to form chains or clusters, creating a sort of mechanical micro-agitation effect. This action probably destroys the early C-S-H bridges and connections between cement particles, further releasing entrained water in agglomerated cement clusters. Therefore, the cementitious suspension shows improved liquid-like behaviour immediately after initiation of the external magnetic field, as depicted in Fig. 3. After a longer period of magnetization, the cement suspension gradually transitions from viscous state to elastic state, most probably due to the improvement of interactions between magnetic clusters and the regeneration of C-S-H bridges between cement particles. After a sufficiently long period of magnetization, the cement paste shows enhanced stiffness compared to the paste without magnetic field owing to the presence of magnetic chains or clusters [58], as presented in Fig. 4. The authors also derived the equations of magnetic force and movement velocity of magnetic particles in cement suspensions, and they also proposed the concept of magnetic yield parameter to describe the relative magnitude of magnetic force and resistance of the suspension [56, 59]. It was revealed that the estimated magnetic yield parameter can be used to predict the magneto-responsive stiffness of cementitious pastes, while the movement velocity has an apparent linear correlation with the early increase in liquid-like behaviour.

The degree of magneto-rheological response of cement paste depends on the paste medium [60, 61], concentration and particle size (magnetic properties) of magnetic particles [57], and magnetic field itself [62, 63]. There exists a suitable solid volume fraction for cement paste with pronounced magneto-rheological response. The magnetic particles are added in a wide range, between 1 and 50% by weight of the cement paste, with the amount used most falling between 1 and 7.5% by weight

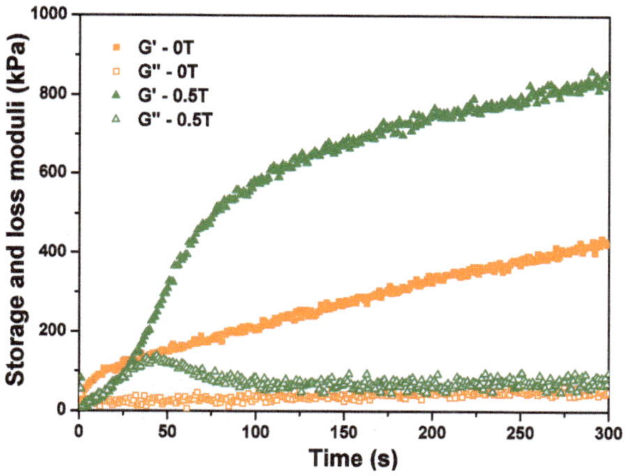

Fig. 3 Typical magneto-rheological response of cementitious paste containing 3% magnetic nano-Fe_3O_4 particles (Courtesy of Dengwu Jiao)

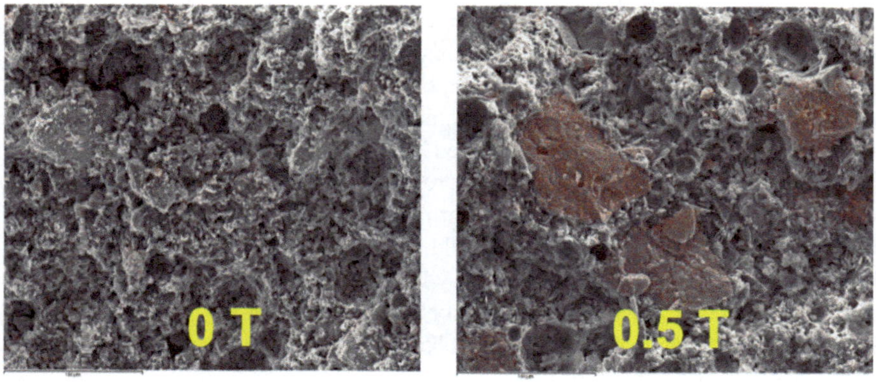

Fig. 4 Representative examples showing the clustering behaviour of nano-Fe_3O_4 particles in cement paste. Red pots indicate the Fe-element (Courtesy of Dengwu Jiao)

of the cement paste [64–70]. Higher magnetic particle content has been shown to enhance stiffening behaviour, likely due to the formation of stronger magnetic particle chains in response to an external magnetic field [55, 57]. With increasing magnetic field strength, the magneto-rheological response becomes more pronounced, with greater amplification and bonding between the magnetic particles [62]. However, an increase in field strength is only reasonable up to a field strength corresponding to the saturation magnetization of the particles, as a further increase does not increase the magnetization of the magnetic particles [70]. With the reduction of magnetic field strength, the structural build-up of cement paste exhibited a sudden drop behaviour [63], due to the reversible behaviour of magnetic clusters in cement suspensions. However, some residual magnetic clusters may be present in the paste after removing the magnetic field, due to the remanent magnetization of magnetic particles and the viscous nature of the suspension, resulting in a higher increase rate of structural evolution [63]. A detailed summary on this subject can be found in [71].

By applying time-varying magnetic fields, it is possible to not only achieve a more solid-like state but also to facilitate a transition from a solid-like to a more liquid-like state. Zhang et al. [72] investigated the effect of time-varying magnetic fields, demonstrating that the shear stress of cementitious materials containing magnetic particles can be reduced on demand. This occurs as the movement of magnetic particles, triggered by changes in the magnetic field, disrupts the flocculated structure. When a bipolar field is applied, the magneto-responsive particles flip, causing a vibration that breaks down the flocculation networks, maintaining the cementitious material in a fluid-like state.

In addition, the dynamic shearing and temperature also exert a significant influence on the rheological response. Jiao et al. [73] found that obvious nanoparticles agglomeration and bleeding were observed on the interface between the paste and the upper rotating plate, owning to the possible nanoparticle migrations under the combined influence of magnetic force and shear rate gradients. They also observed that limestone suspension containing magnetic particles showed an immediate enhancement in stiffness upon applying a stepwise-increased magnetic field [74], due to the fact that the formation of magnetic clusters exerts an immediate resistance to the shear flow. Despite the disruption of the chains or clusters by the shearing, new clusters are simultaneously formed, as the increased movement of particles raises the likelihood of magnetic particles encountering each other and creating new or larger clusters. A stepwise decreasing magnetic field resulted in a drop in shear viscosity, followed by a gradual further reduction. In addition, the residual magnetic clusters can be possibly completely destroyed by the low-rate shearing [74], leading to a comparable apparent viscosity like those without magnetic field. Recently, Jiao et al. [75] observed that the rheological response of cement paste with magnetic nano-Fe_3O_4 particles drops significantly when the temperature falls below 20 °C (by 4–8%) or exceeds 30 °C (by 25–47%). This is mainly due to the change of the hydration rate of the cement paste under varying temperatures.

While CIP and nano-Fe_3O_4 particles are effective at providing magnetorheological properties to cementitious materials, their high cost presents a barrier to widespread application even when used in small quantities [76]. Exploring the use of affordable, magnetically active cement replacement materials appears to be a viable and cost-effective alternative. As a result, recent studies have explored using affordable cement replacement materials with magnetic properties, such as fayalite slag, waste iron slags, and fly ash as alternatives [77–81]. These studies indicate promising potential for these alternatives to provide magnetorheological properties, though effectiveness varies depending on the type and amount used. Nevertheless, further research into optimizing the magnetic properties and quantities of these alternative materials is required to advance the practical application of sustainable MR control in construction materials.

As aforementioned, magnetic particles with low remanent properties are beneficial for applications requiring reversible control. However, certain applications benefit from high magnetic remanence, where stability is needed after magnetic activation, such as 3D printing, well cementing, and formwork leakage prevention. Jiao et al. [82] explored this in the context of 3D concrete printing through rheometric tests, using magnetorheological control to improve the material's buildability by applying a pulsed magnetic field just before extrusion for the mixtures containing two types of nano-Fe_3O_4 with different remanent magnetizations. They observed improved structural build-up with the application of the pulsed magnetic field, regardless of pulse duration of 1 and 5 s, and greater structural build-up was observed for the paste including the powder exhibiting higher remanent magnetization. Additionally, hard magnetic particles, which retain their magnetism after being magnetized [83], may have potential to achieve a more permanent increase in rheological properties after removing the magnetic field. By mixing hard magnetic particles in the cementitious materials, a new type of magneto-responsive cementitious material was developed, which can sustain a certain rheological and magnetization response after the magnetic field was removed [84]. Figure 5 show a typical example of the alignment of neodymium iron boron ($Nd_2Fe_{14}B$) in cement paste after removing the magnetic intervention.

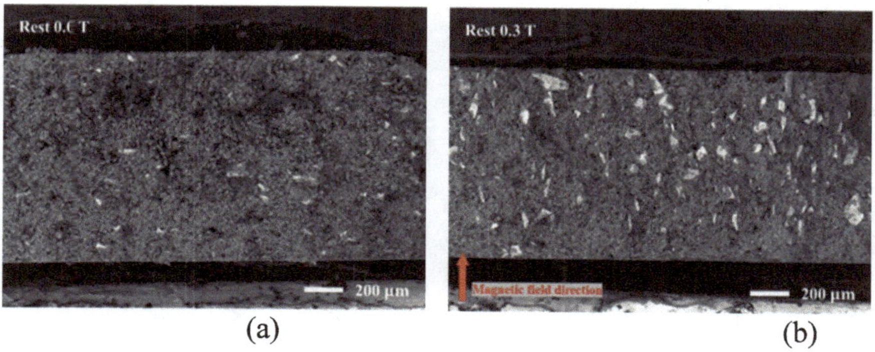

Fig. 5 Morphology of responsive cementitious materials **a** before and **b** after a magnetic intervention of 180 s at 0.3 T. (Courtesy of Yiyuan Zhang)

The active rheology control of cementitious materials through magnetic field application is still in its initial stage and quite innovative. Most publications in this field have been conducted at the cement paste level. Well cementing is one of the application areas that stands to benefit from this technology at the paste level. Kresinski et al. [64] demonstrated that magnetorheological cement paste could provide effective pressure sealing for plugging and abandonment operations in oil and gas wells. Their research showed that applying magnetic fields in the range of 0.1 T activated the cement paste, enhancing its ability to form a secure barrier for plug setting within the wellbore or annular space. This offers a potential solution to the challenges of well cementing, where rheological control is critical for successful cement plug placement.

On the other hand, to make active rheology control broadly applicable at construction site, an upscaling to at least the mortar scale is required. This was firstly undertaken by Zhang et al. [85] who found that with increasing sand content the magneto-rheological response decreases in terms of percent increase of yield stress and response speed as the sand particles own a higher resistance force against the movement of the magnetic particles as the cement particles. Meanwhile, the alignments of magnetic particles in the cement mortar were captured and visualized by SEM (scanning electron microscope) in this research. Figure 6 presents the distribution of magneto-responsive particles around the sand particles.

In order to further improve the magneto-response of cementitious materials, magneto-responsive aggregates (MRAs) are applied to develop an improved responsive mortar by Zhang et al. [86]. A sound experimental validation, qualitative analysis and quantitative characterization of magnetorheological response and mechanism of cementitious materials containing responsive aggregates under a nozzle/pipe-like (inline) external magnetic field is established. After applying the magnetic intervention with the gap of several centimetres, the yield stress determined by penetration tests and torque determined by shear tests increased significantly. Meanwhile, the alignments of magneto-responsive aggregates are visualized. X-ray CT images of cement mortar containing MRAs with and without the intervention of a magnetic field is shown in Fig. 7.

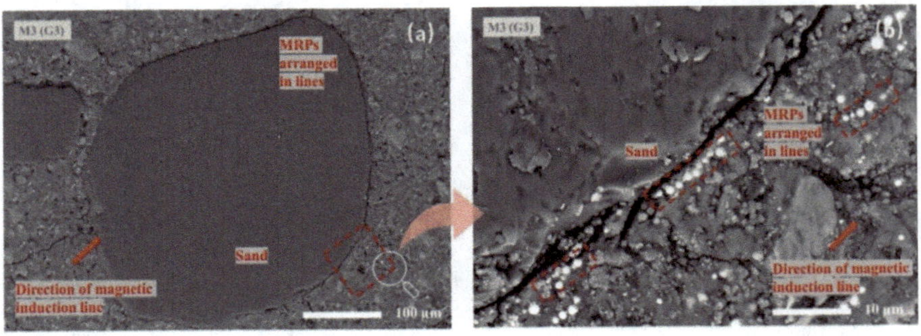

Fig. 6 Distribution of magneto-responsive particles around the sand particles [86]

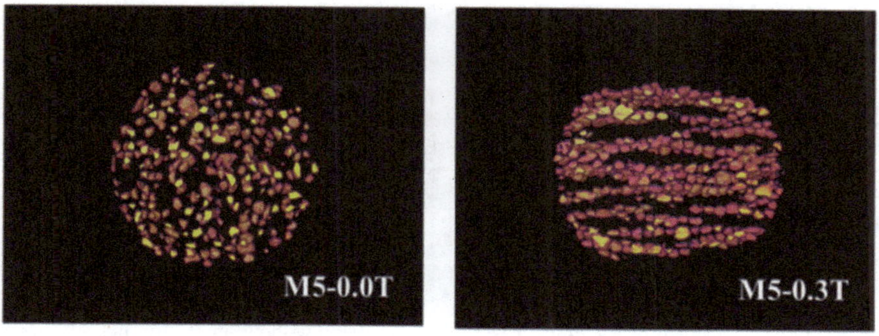

Fig. 7 X-ray CT images of cement mortar containing MRAs with and without the intervention of a magnetic field [87]

Recently, Guo et al. [87] explored the magneto-rheological behaviour of cement paste using steel fibres as responsive elements. It was found that after experiencing a magnetic field, the flow velocity decreased and the flow diameter was smaller in the direction parallel to the field compared to the perpendicular direction, likely due to the flow-restricting effect of residual fibre aggregations formed under the magnetic field. These findings provide fundamental experimental support for actively controlling the rheological properties of steel fibre-reinforced cement-based materials. Hu et al. [88, 89] found that magnetic field has no clear influence on the early-age hydration products, while the movement of magnetic particles contributes to the formation of weak interfacial zones between magnetic particles and the matrix, resulting in a mechanical anisotropy behaviour. Moreover, with regard to the potential application of magneto-rheology control in 3D concrete printing (3DCP), it is important to determine the optimal location for applying the magnetic field. Two potential scenarios are the application at the print-head or at the print-bed, with each differing in terms of the timing of field application. Kanagasuntharam et al. [69] investigated both approaches and found a significant difference in static yield strength. Continuous application at the print-bed resulted in much higher static yield strength compared to the brief application at the print-head. Technical feasibility has to be considered as well.

3.2 *Electrical-Induced Rheology Control*

Electro-rheology refers to the alteration of the rheology of fluids by the application of electric fields, as first reported by Winslow in 1947 [90]. Sometimes, this phenomenon is called the "Winslow effect" [91, 92]. There are several strategies to use electric field to achieve rheology control. Conventional electro-rheological (ER) fluids are colloidal suspensions whose properties depend strongly and reversibly on the electric field [91]. The viscosity of the suspension fluids can increase by factors

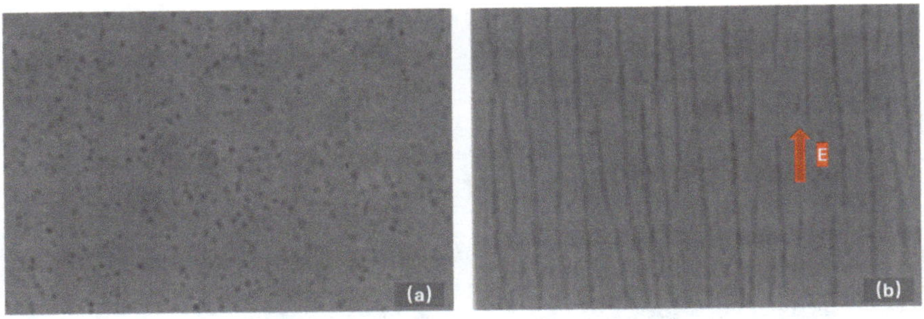

Fig. 8 Basic mechanism of ER fluids **a** Isotropic suspension without an electric field; **b** Fibrillated structures with an electric field (redrawn after J. E. Martin and J. Odinek, Sandia National Laboratories) [91]

as high as 10^5 after applying an electric field. Figure 8 presents the basic mechanism of ER fluids, which is the most common and traditional method. With the intervention of an electric field, fibrous structures are formed along the applied field, resulting in increase in yield stress and viscosity. When using this electrical-induced control strategy, the basic mechanism is similar to that of MR fluids. ER fluids are technically more challenging, due to sensitivity to impurities and temperature, and relatively high, general voltage demands [93]. This approach is not common in the field of cement-based materials.

The first approach to apply an electric field to improve rheology of cementitious materials was performed by Choi et al. [94]. In a pumping pilot test, they applied an externally imposed electric field on the carbon steel pipe in order to improve the formation and/or the rheological properties of the lubrication layer. This technique has been inspired by a previously developed formwork removal method based on electro-osmosis [95, 96]. Although it was experimentally shown that the electric field applied by Choi et al. [94] reduced the required pumping pressures, the fundamental mechanism has not been fully revealed up to now. It might be that the imposed electric field did not really affect the rheological properties of the lubrication layer, nor of the bulk concrete. Because of the use of a carbon steel pipe, any electromagnetic field would have been significantly shielded, not reaching the pumped concrete inside the pipe. The real mechanism might have been the effect of the electric field on water molecules near the pipe surface, forming a thin layer of migrated polarized water molecules acting as a slip layer. In this way, the concrete itself was not rheologically improved by the electric field. As there was no electro-responsive component in the concrete (except for the possible polarization of water molecules), they still passively relied on the rheological properties of the pumped concrete.

In a fundamentally new approach, the SmartCast research team of Ghent University, Belgium, introduced the application of electro-responsive polymers in cementitious materials in view of actively controlling the rheological behaviour [19]. Therefore, and contrary to what has been done by Choi et al. [94], this polymer fraction is an indispensable element in the active rheology control mechanisms, responding to an external impulse, i.e. the electric signal, inducing a change in rheological properties of the cementitious material. While different options have been considered, including cleavage of side chains, the most pronounced results have been obtained with redox-responsive polycarboxylate ethers with TEMPO side chains [97, 98]. Figure 9 gives a schematical drawing of the newly developed switchable superplasticizer, consisting of a classical backbone with negative charges (MAA) and of side chains of a new type, namely TEMPO [97]. This side chain on the one hand can provide steric hindrance, and on the other hand can be triggered by an electric signal to become positively charged. Upon electric triggering, the positive charges of TEMPO+ will counterbalance negative charges from the backbone, reducing the adsorption capacity to the cement grain surface. In this way, a controllable superplasticizer is obtained, enabling active rheology control by adjusting the adsorption capacity through external electric redox signals.

The controllable rheology of the cementitious material containing the redox-responsive superplasticizer is illustrated in Fig. 10 [99], showing flow curves obtained with plate-plate rheometry for the reference paste as well as for the controllable pastes containing 0.1% (by weight of cement) of the switchable TEMPO-admixture. In this experiment, the different control states of pTEMPO were obtained by chemical oxidation. However, as illustrated further, this can also be achieved by redox control. All pastes had a water/cement ratio equal to 0.35. Viscosity and yield stress

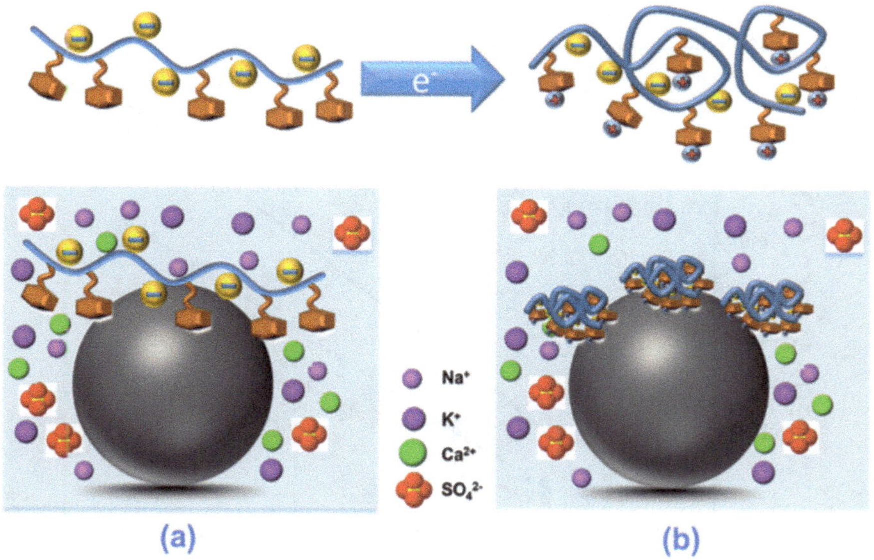

Fig. 9 Schematic representation of the mode of action **a** poly(MAA-co-TEMPO) shows an extended polymer chain adsorbed onto cement particles, and **b** poly(MAA-co-TEMPO+) forms coils due to the intramolecular interactions between the positive and negative charges in the polymer chains causing dehydration and aggregation (For reasons of clarity, cement particles and polymers are not drawn at the same scale) [99]

values have been obtained by fitting a linear Bingham model to the experimentally obtained datapoints. The reference cement paste without admixture showed yield stress and plastic viscosity of about 35 Pa and 1.36 Pa s, respectively. In case of addition of pTEMPO the yield stress and plastic viscosity lowered to 20 Pa and 0.87 Pa s, respectively, showing better fluidity of the cement pastes. The addition of chemically oxidized pTEMPO+ resulted in increased yield stress and plastic viscosity compared to pTEMPO, as could be expected based on the previously illustrated working principle of the TEMPO-based polycarboxylate ether superplasticizer.

The effect of the switchable superplasticizer on structuration was monitored by means of SAOS tests, for pastes containing TEMPO radical as a redox responsive moiety, before and after the application of an external voltage trigger. The voltage was applied via a potentiostat connected to a specially developed screen-printed electrode (SPE) positioned on the bottom plate of the rheometer. More details on this setup can be found in literature [97, 99]. The obtained results are presented in Fig. 11, showing the storage modulus (G′) of two different cement pastes with and without electric trigger, namely one reference cement paste containing a lab-synthesized non-responsive (Polycarboxylate) PCE superplasticizer, and one paste containing the newly developed TEMPO-based switchable PCE superplasticizer. Without any trigger signal, both cement pastes showed a comparable evolution of the storage modulus G′. Both pastes were also exposed to a trigger of 0.85 V at 75 s after the start of the experiment. This trigger was further maintained until the end of

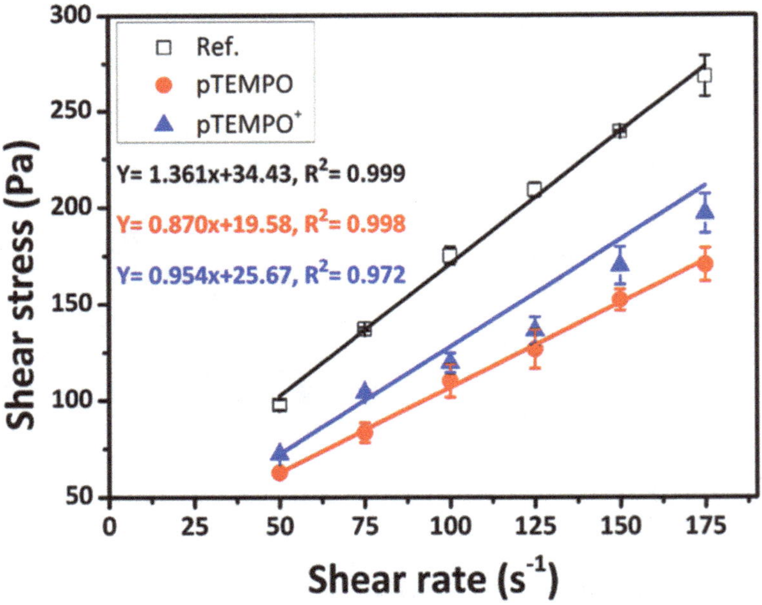

Fig. 10 Flow curves of the cement pastes with copolymers before and after chemical oxidation (polymer concentration = 0.1 wt.% and w/c = 0.35) [98]

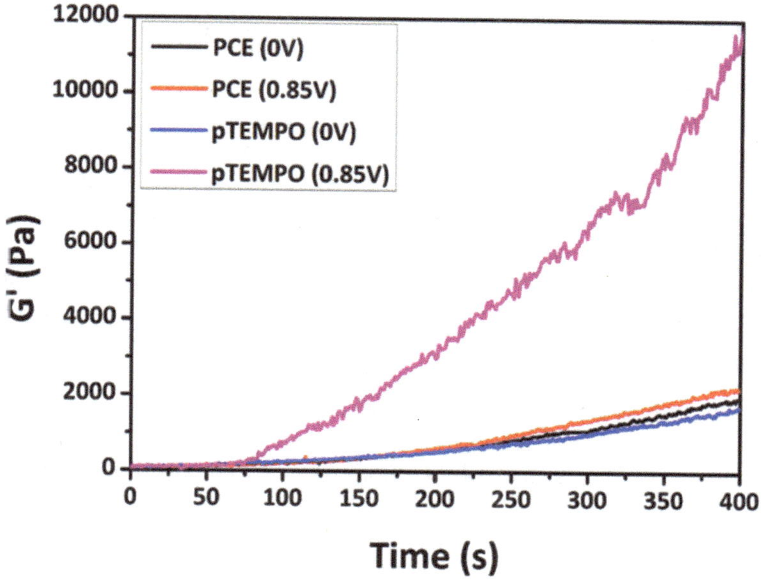

Fig. 11 Storage modulus of cement pastes containing 1 wt.% pTEMPO before and after applying an electric potential of 0.85 V at 75 s, compared to reference cement paste (W/C = 0.35) containing a lab-synthesized PCE without responsive unit [98]

the experiment. With trigger signal after 75 s, the cement pastes with the TEMPO-based controllable admixture showed a rapid increase in storage modulus G′, while the classical PCE-based system was not influenced by the trigger signal. The huge effect noticed within the controllable cement paste confirms that redox-responsive superplasticizers could be considered to achieve active stiffening control (ASC).

3.3 Vibration-Induced Rheology Control

Concrete vibration is a well-established practice in the construction industry, commonly used to compact concrete into formwork during casting. Concrete exhibits thixotropic behaviour, characterized by a reduction in viscosity under shear, which in this case is induced by vibration. Concrete exhibits an immediate reaction to vibrations; under such conditions, its yield stress diminishes significantly, occasionally reaching zero [100]. Similar observations are also noted by Tattersall et al. [101]. In his other study, the authors reported that vibration could reduce the yield stress to zero, causing the concrete to behave like a Newtonian fluid [102], and can flow due to its own weight. Generally, the response of concrete to vibration depends on the amplitude surpassing a minimum threshold and staying below a maximum frequency limit [102]. Further, Li et al. [103] stated that vibration must not be intense or prolonged to the extent that it causes segregation of concrete.

Building on this understanding, Sanjayan et al. equipped a print head with a vibrator to enable active control over the extrudability of the concrete, which is primarily governed by its rheological properties (see Fig. 12) [104]. This study found that the application of vibration not only improved the extrudability of the concrete but also enhanced its buildability. The improved extrudability resulted in layers with reduced deformation and fewer interlayer defects, thereby enhancing the structural integrity. Notably, the influence of vibration on the rheological properties of concrete remains minimal until a critical vibration velocity is achieved, which is dependent on the specific mix composition. However, the intensity and duration of the vibration must be carefully controlled to prevent excessive reduction in the rheological properties of the constitutive paste, as this could compromise its ability to resist aggregate sedimentation under gravitational forces, causing segregation in the concrete. While the study [104] employed mechanical vibrators applied to the external surface to control the rheological properties, prior research has explored technologies for inducing internal vibration by incorporating magneto-responsive particles into the concrete mix. This approach is detailed in Sect. 3.1.

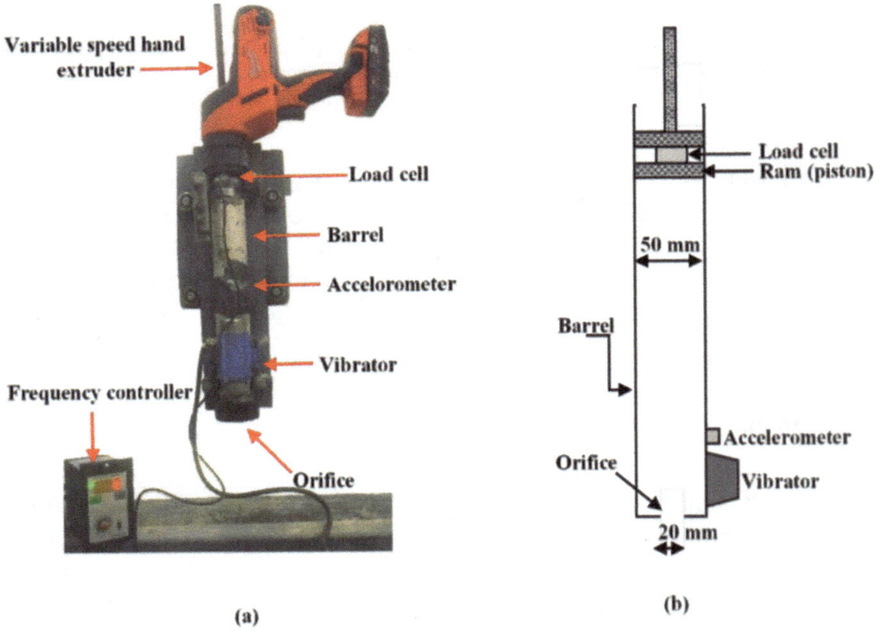

Fig. 12 **a** Photograph of the vibration-extrusion setup and **b** Schematic representation of the vibration-extrusion setup [104]

3.4 Temperature-Induced Rheology Control

Temperature-induced rheology control can be divided into two approaches, direct control and indirect control. Direct control is primarily inspired by the temperature effect on hydration mechanism of cementitious materials, which involves the immediate effects of temperature on the properties of raw materials and the hydration kinetics. An increase in temperature reduces the viscosity of water, facilitating improved flowability of cement-based materials [105]. This phenomenon is more pronounced in mixtures with a high water-to-cement (w/c) ratio [106]. As the temperature increases, the evaporation loss of water enhances as well, resulting in a stronger friction between particles, which in return increases the yield stress and viscosity of materials [106, 107]. Additionally, elevated temperatures enhance the Brownian motion of cement particles, especially for those lower than several micrometres, leading to an increase in the probability of collision between cement particles, therefore accelerating cement hydration and increasing yield stress and viscosity of cement-based materials [108, 109]. Furthermore, temperature can accelerate the dissolution of cement particles, which is beneficial to cement hydration. Muthukrishnan et al. [110] demonstrated the potential of microwave heating for rheology control of geopolymer mortar. By applying microwave heating with a power of 1200 W for 10 s, they accelerated the hydration of geopolymer, resulting in a substantial increase in the average Young's modulus of geopolymer mortar from 290 to 2047 kPa. To improve the rheological adaptability of cement-based materials to temperature variations, Qu et al. [111] introduced a thermal responsive gelatine in cement paste. Their study revealed that the modified paste exhibited a yield stress of 1.9 kPa after resting for 10 min at 5 °C, which was 10 times higher than that of the reference paste. Similarly, Christ et al. [112] adopted hydrogel as binders to make temperature-sensitive polymer-based concrete. This concrete remains fluid with a proper viscosity at elevated temperatures. As the temperature decreases, both the viscosity and yield stress increase. The concrete hardens when the temperature falls below the gel point of the hydrogel binders.

Indirect rheology control by temperature involves utilizing temperature as a trigger to activate specific components within the cement-based material that influence its rheological properties. Ramakrishnan et al. [113] and Shao et al. [114] explored the use of thermosensitive capsules that melt upon heating, thereby releasing chemical admixtures encapsulated within, enabling targeted modification of the rheological properties of cement-based materials. Zhang et al. [109], inspired by the influences of polymers on the temperature sensitivity of rheological properties, proposed an approach of active rheology control for cement-based materials coupling temperature control with the utilization of viscosity modifiers (Fig. 13). They found that increasing temperature improves the efficacy of hydroxypropyl methylcellulose (HPMC) and polycarboxylate ether-based superplasticizer through accelerated dissolution and adsorption, and enhanced chain stretching. As a result, it allows for an intensified or a suppressed response to temperature in the apparent viscosity, thixotropic area, static yield stress and storage modulus.

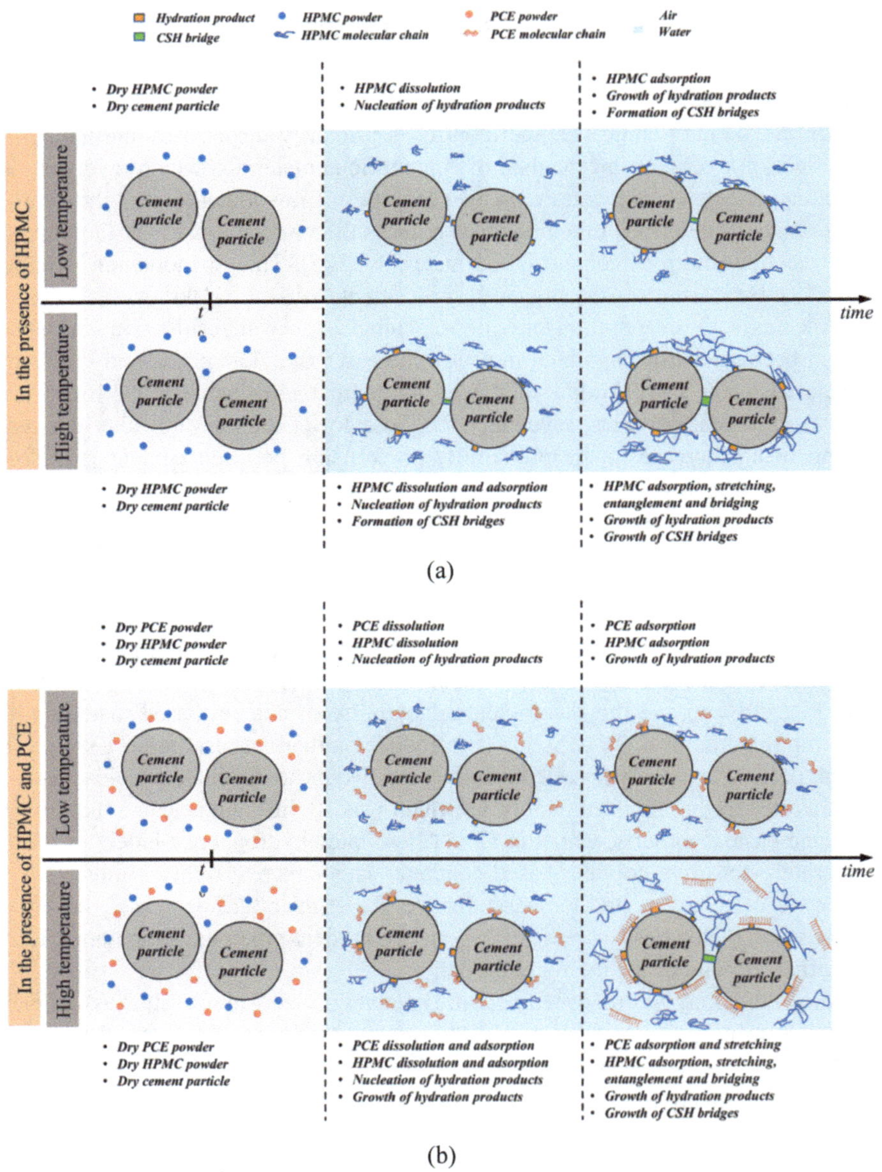

Fig. 13 Interaction mechanisms between polymers and cement particles under temperature variations: **a** HPMC; **b** HPMC and PCE [109]

3.5 *Microwave-Induced Rheology Control*

Microwave-induced rheology control is mainly the extension of temperature-induced rheology control. There are several heating techniques available, but compared to others, this technique provides higher heating rates along with uniform heating. Microwave heating is generated by dissipating the internal energy associated with excited molecules of dielectric material kept in an electric field. According to [18], these are basically electromagnetic waves with high frequencies (300 MHz–300 GHz) and wavelengths ranging from 1 mm to 1 m. A few studies [115, 116] exist regarding the curing of concrete structures through microwave heating. However, there is limited research available on the control of rheology.

Muthukrishnan et al. [110] found that the forces required for the agglomeration of polymers, such as Van der Waals forces or other electrostatic forces, are significantly weaker in geopolymer binders, resulting in a lack of structuration comparable to cementitious materials. Thus, it will affect the printability characteristics of the geopolymer-based printable concrete. Therefore, the author employs microwave heating to improve interlayer bonding and buildability in geopolymer-based printable concrete. In his study, he used the 10s microwave heating to check the interlayer bonding. However, at the same time, to maintain the similar properties of printed materials, he found the duration of microwave heating for the viscosity recovery test is 33 s.

The protocol used for the viscosity recovery test is shown in Fig. 14. In this one, stage 1 of 60 s mimics the condition before the extrusion, stage 2 of 30 s mimics the stage during the extrusion and stage 3 of 60 s, similar to stage 1, i.e. after the extrusion stage. The authors have applied microwave heating after the extrusion to check the interlayer bond strength. Therefore, to maintain the similarity, authors kept the 40 s resting period between stage 2 and stage 3 for the control mix while providing heating during that interval for the other mixes. The result for the viscosity recovery test is shown in Fig. 15. Authors noted that microwave heating significantly alters the reconstruction of the fresh mixture following pumping or extrusion. All the mixtures in stage 1 exhibit viscosity values in the range of 10^{-5} Pa, which are reduced by 10^{-4} Pa during the extrusion stage. At stage 3, the control mix recovers up to 25% of the initial viscosity, while the microwave-heated mixes demonstrate a complete 100% viscosity recovery. This enhancement in a viscosity recovery indicates the capacity of geopolymer mixes with high-quality printing properties. Prior to utilizing microwave heating for the specific mixture, it is essential to determine the appropriate heating duration. Excessive heating of the mixture accelerates hydration, resulting in a very stiff mix that adversely impacts other printing properties.

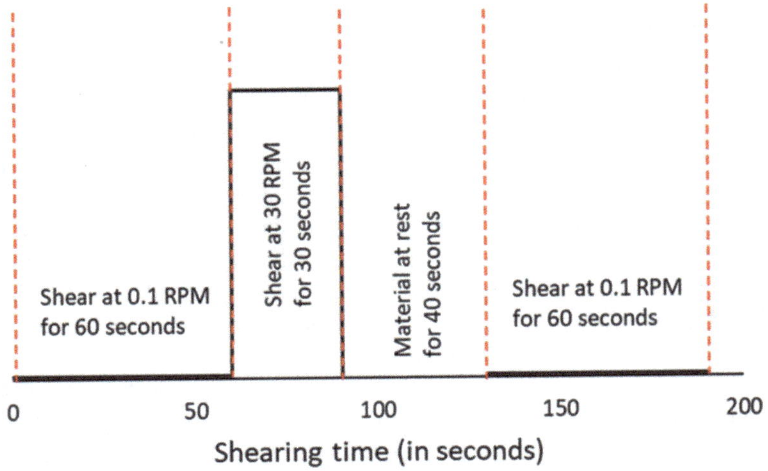

Fig. 14 Shearing protocol used for the viscosity recovery test [110]

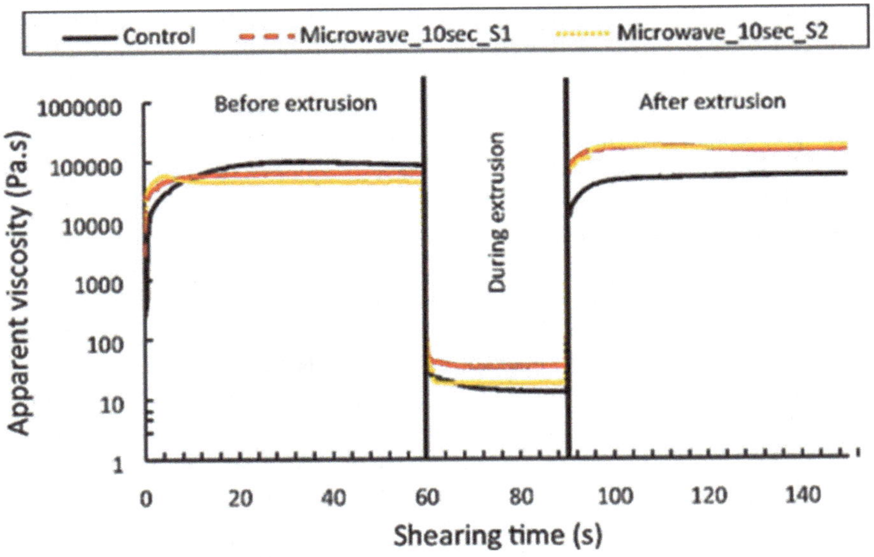

Fig. 15 Viscosity recovery of mix before and after microwave heating [110]

3.6 Ultrasonic-Induced Rheology Control

Application of ultrasound is a well-established technology applied during processes in view of delivering high-quality end products in significantly less time and with reduced energy consumption [117]. The high pressure and temperature associated with the intervention process is expected to induce changes in rheological properties

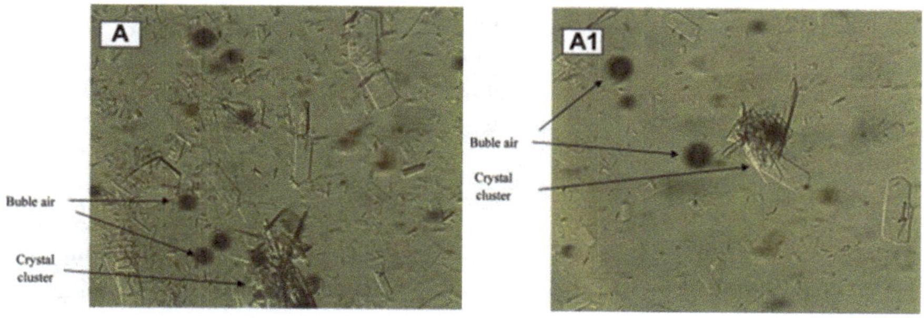

Fig. 16 Crystallization of honey before and after ultrasonic treatment **a** 40 °C Heat-treated samples after 20 min of thermal treatment; **a1** 40 °C and ultrasound-treated samples after 20 min of treatment

[117]. In the food industry, it is used commonly to change the rheological properties. For example, ultrasound treatment speeds up the liquefaction of honey [118]. As shown in Fig. 16, by destroying the crystallization, the fluidity of honey increases after ultrasound treatment. In addition, the ultrasound treatment is also used to decrease the viscosity of proteins including myofibrillar proteins [119], faba bean proteins [120], etc.

In cementitious materials, ultrasonic signals are used to unlock agglomerates of particles and homogeneously distribute the fine particles efficiently at the micrometric and nanometric scale, with different morphologies and compositions, that tend to aggregate and flocculate. The mechanism of ultrasound dispersion is mainly based on acoustic cavitation, referring to the formation, growth and implosion of bubbles [121]. The ultrasonic dispersion mechanism is further broken down by Xu et al. [122], who divided the ultrasonic influences according to their dispersion effects caused by mechanical action and cavitation and the ultrasonic enhancement. The latter results from an accelerated dissolution of the cement particles and a promoted microstructure formation. Several authors showed advantageous effects of applying ultrasonic waves to cement pastes by reducing the apparent viscosity and the initial yield stress.

Studying cementitious materials containing silica fume, Wang et al. [123] applied sonication (during 12 min) to the silica fume in aqueous solution prior to adding it to the cement paste. The workability (spread) of the cement paste containing sonicated silica fume was lower than in the reference case without sonication. The reason is attributed to the higher specific surface in case of sonicated silica fume, requiring more water for wetting. Explained in more scientific details, Wang et al. [123] refer to the formation of the double electric layer near the surface of the silica fume, adsorbing more water on the charged surfaces. This leads to a reduced amount of free water in the paste, reducing workability. If the sonication can be done directly within the cement paste, this methodology could be considered as active control of the rheological properties.

Silvestro et al. [124] even went one step further, also studying the effect of sonication on polycarboxylate-based superplasticizers in a cementitious matrix. Rheological tests were performed with plate-plate rheometry with a gap width of 1 mm. The details of the test protocol can be found in [124]. The data of the obtained flow curves were fitted based on a Herschel-Bulkley model, resulting in values for yield stress and viscosity. Thixotropy was quantified by calculating the area between the ascending and descending curves. Sonication of the superplasticizer resulted in increased yield stress, viscosity, and thixotropy. By means of more in-depth tests, including dynamic light scattering, they concluded that the sonication led to a decrease in chain length of the PCE superplasticizer, reducing its dispersing capacity. The reduced dispersing capacity further reduced the flowability of the cement paste (higher yield stress and viscosity). An effect on the compressive strength of the pastes at 28 days was not noticed. The research of Silvestro et al. [125] suggests a potential control mechanism for the rheology of cement paste, by controlling the adsorption capacity of PCE on the cement grains, triggered by the effect of sonication on the polymer chain length. Ultrasonication was also applied to functionalize carbon nanotubes (CNTs) in Portland cement pastes Silvestro et al. [124]. An improved dispersion of different graphene additions could also be obtained by Zou et al. [126] and Liu et al. [127].

De La Rosa et al. [128] investigated pastes composed of 100% type I Portland cement, as well as blends containing 80% type I Portland cement combined with 20% metakaolin or 20% natural volcanic pozzolan. The application of ultrasound significantly reduced thixotropy in all cases, with the most pronounced effect observed in the 100% Portland cement paste. Ultrasonic treatment was applied for 1 min using a Hielscher ultrasonic device UP400S, operating at a maximum power of 400 W. The water-to-cementitious materials ratio (w/cm) was fixed at 0.25, and the superplasticizer-to-cementitious materials ratio (SP/cm) was maintained at 0.02. Rheological properties of each suspension were evaluated using a rotational rheometer equipped with a plate–plate sensor (35 mm diameter, 2 mm gap). This configuration ensures the sample completely fills the measurement volume and maintains steady laminar flow. After mixing each suspension, both with and without ultrasonic treatment, samples were tested at a controlled temperature of 25 ± 2 °C. Shear rate was carefully controlled to prevent slippage between the suspension and the sensor walls. The rheological test protocol consisted of a gradual increase in shear rate from 0 to 200 s^{-1} over 60 s, holding the maximum shear rate for 30 s, followed by a gradual decrease from 200 to 0 s^{-1} over 60 s. This procedure ensured a uniform and fully de-agglomerated mixture.

3.7 PH- and Chemical-Induced Rheology Control

The pH- and chemical compositions also have the potential to control the rheology of cementitious materials [129]. Cementitious materials typically have a high pH, ranging from 12 to 13, indicating a basic chemical environment. Adjusting the pH can be used to control the rheology of cementitious materials, because it affects the surface charge and interactions between the suspended cement particles. Adding alkaline admixtures like sodium hydroxide increases the pH, enhancing the negative surface charge on cement particles. This leads to better dispersion and lower viscosity. The study in [130] observed two different apparent viscosity curves at different pH values, suggesting that specific pH ranges might have unique effects on the viscosity behaviour. However, further research is needed to fully understand these specific pH-dependent effects. At different temperatures, pH value may also have different effects on rheological properties. Some research reports that higher pH (from 7 to 13) increased the yield stress of cement at temperatures below 30 °C, which was attributed to the enhanced cement hydration as a result of higher pH, producing more hydration products and a denser microstructure [130]. However, at higher temperatures (above 40 °C), a higher pH led to a reduction in yield stress, as it was speculated that the increased ion activity [anions (OH^-) and cations (Na^+)] and hydration acceleration led to breakdown of the microstructure, resulting in lower yield stress [130].

Based on this principle, Kanagsuntharam et al. [113, 131] investigated the use of encapsulated sodium hydroxide and sodium silicate to achieve on-demand setting in 3D-printable concrete mixes. These encapsulated alkaline substances were incorporated during the wet mixing stage of the concrete preparation. Subsequently, the concrete was heated at the print head to dissolve the capsules, releasing the alkaline substances to modify its rheology. The study found that increasing the dosage of the alkaline activators led to a rise in static yield stress. However, the hardened properties, specifically the 1-day and 7-day compressive strengths, decreased as the dosage of alkaline substances increased. It was hypothesized that the higher pH accelerated the dissolution of aluminate compounds from the cement particles, promoting the rapid formation of ettringite, which significantly increased the yield stress. However, the enhanced formation of ettringite caused sulphate depletion in the mix, reducing the formation of C-S-H. This reduction in C-S-H formation was identified as the primary factor contributing to the diminished hardened properties of the concrete.

The pH also plays a crucial role in regulating the rheology of alternative binders to Portland cement, such as geopolymers. For sodium hydroxide and sodium silicate-activated geopolymers, an increase in pH enhances the rheological properties of the concrete. However, beyond a certain pH threshold, the hardened properties of the geopolymers begin to decline, highlighting the importance of defining the pH limits to achieve effective rheological control. The silicate modulus, defined as the ratio of SiO_2/Na_2O, is usually adjusted by varying the ratio of sodium silicate to sodium hydroxide in the mix [132]. Therefore, the mixes with a lower silicate modulus exhibit higher pH due to a greater proportion of sodium hydroxide. The early-age rheological properties, characterized by the spread diameter, were found to increase as the pH of the activator solution decreased, except at a silicate modulus (Ms) of 2.15 (Fig. 17). Most likely, at this silicate modulus, the accelerated formation of C-S-H was facilitated by the abundant availability of silicate anions, thereby negating the expected reduction in rheological properties associated with a lower pH. For the hardened properties, characterized by 7- and 28-day compressive strength, a reduction in the pH of the activator solution led to an increase in compressive strength, where a significant drop in strength was observed when the silicate modulus was reduced from 1.04 to 0.54.

Chemical intervention is currently the most widely used method to improve the rheological properties of cementitious materials. On the one hand, with the vigorous development of the chemical industry, many highly effective admixtures have been invented and widely used, such as superplasticizers, retarders, accelerators, viscoelastic modifiers, etc. For example, concrete superplasticizers disperse cement particles through electrostatic repulsion, steric hindrance and wetting, reducing water consumption and improving fluidity, strength and durability. Retarders delay cement hydration reaction, extend setting time and improve construction performance by adsorbing cement particles, forming complexes or reducing solubility. Cement accelerators accelerate coagulation and hardening and improve early strength by promoting rapid hydration of C_3A and C_3S, forming early hydration products or adjusting the pH value of the solution. On the other hand, in some fields such as 3D printing concrete, a two-component (2K) system is also developed. The 2K system in 3D printing concrete refers to the use of two-component printing materials to optimize the performance and process of cement-based 3D printing. It usually involves two independent material streams that mix or react in real time during the printing process to improve printability, mechanical properties and curing speed.

3.8 Pressure-Induced Rheology Control

Pressure can influence the rheological behaviour of cementitious materials. This is clearly observed in the case of concrete pumping [6]. The pressure-induced high shearing energy can lead to a better dispersion of cementitious particles [133, 134], leading to a decrease in viscosity during pumping. The yield stress can increase, remain constant, or decrease during pumping [134]. In addition, the air-void system in the concrete will be affected by pumping under pressure. The air content can increase, decrease or remain constant during pumping under pressure [135, 136]. The capillary number, defined as the ratio of the applied shear stresses to the surface tension of air bubbles, can help to explain the effect of pressure-induced shear stresses on the rheological properties of the pumped concrete containing air bubbles [137].

Based on the pumping observations, pressure could be further considered as an active way of controlling rheological behaviour of cementitious materials. Pressure-induced control is commonly used in practical applications. For instance, to increase the flow rate during the pumping process, the higher pumping pressure was used to enhance the shear thinning behaviour of cement-based materials [134]. In the application of extruded concrete, the reduction in the cross-section of the nozzle can compensate for the decreased pumping pressure during pumping. The improved pumping pressure leads to a decrease in viscosity of the materials at the nozzle, therefore densifying the materials and ensuring a smooth surface of the extruded materials [138]. The pressure-based compaction method is usually adopted for better packing and stiffening of the cementitious material.

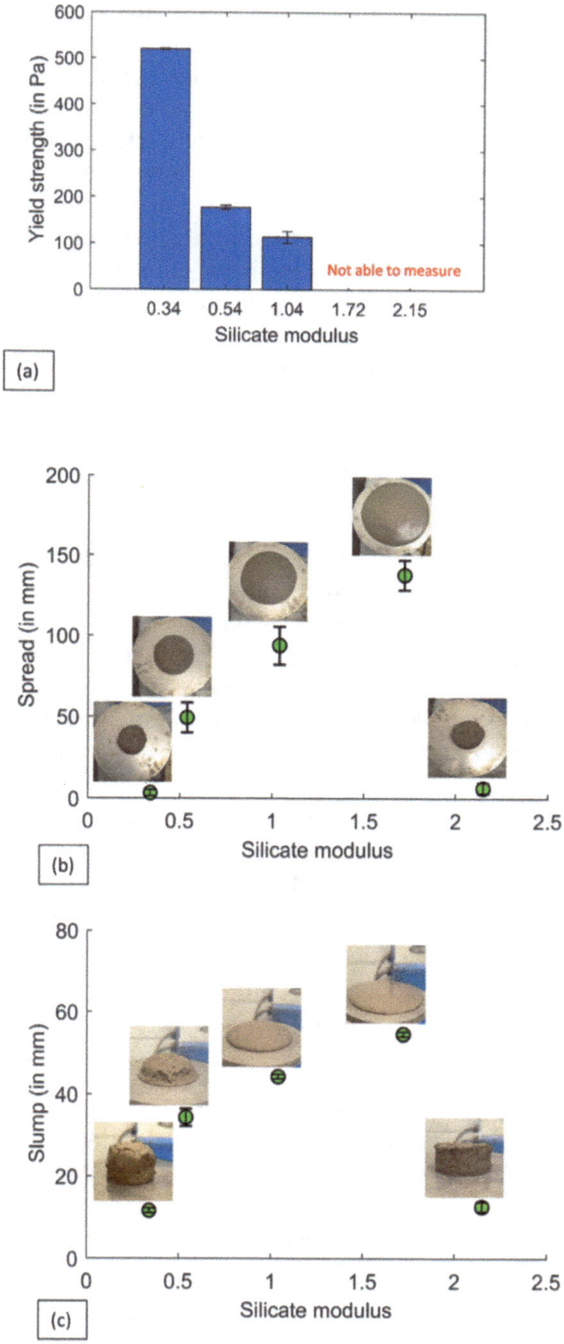

Fig. 17 Influence of the activator's silicate modulus on **a** dynamic yield stress, **b** spread diameter, and **c** slump of the alkali-activated mix. The figure is reproduced from [133]

High pressure in closed systems significantly affects the hydration and rheological behaviour of cementitious materials. Early studies showed that pressure accelerates cement hydration, increasing the solid volume fraction and viscosity, which reduces handling time but complicates applications like deep-well cementing [139]. In cement pastes, hydrostatic pressure interacts with admixtures, where superplasticizers reduce yield stress, but silica fume and nano-clay increase it due to water adsorption, altering the flowability of mixtures under high-pressure conditions [140]. At lower water-to-cement ratios, high pressure can reverse thixotropic behaviour into rheopexy, stiffening the material under shear, whereas such effects are negligible at higher ratios [141]. In addition to influencing rheology, hydration under pressure transforms free pore water into gel water, increasing static gel strength and refining pore structures. This transition reduces pressure transfer in cement slurries and occurs through a two-stage process: initial slow decay from particle bonding followed by rapid reduction from hydration products [142]. Elevated pressure also promotes the formation of hydration products like ettringite and rodlike gypsum, which increase the specific surface area and reduce particle separation, significantly raising yield stress in cement pastes [143]. These phenomena collectively illustrate how high-pressure conditions alter hydration dynamics, microstructure, and rheological behaviour, offering insights critical for optimizing processes such as deep-well cementing and concrete pumping.

In summary, high shear energy enhances particle dispersion, reduces viscosity, and improves the flow properties of concrete during pumping. Conversely, in sealed systems, pressure accelerates cement hydration, increasing the formation of hydration products and solid volume fraction. This results in higher viscosity as the material transitions to a more rigid structure. These observations suggest that pressure could be explored as an active method to control the rheological behaviour of cementitious materials. However, to date, no comprehensive studies or practical applications have demonstrated active rheology control through pressure, aside from pressure-based compaction techniques, which improve particle packing and material stiffening.

3.9 Prospects and Summary

This chapter has comprehensively reviewed the recent advances in the active control of rheological properties of fresh cementitious materials. The discussion began with fundamental rheological concepts, including viscoelasticity, yield stress, plastic viscosity, thixotropy, shear thinning, and shear thickening, which are essential for understanding the flow behaviour and workability of cement-based materials.

Various active rheology control methods, such as magnetic fields, electric fields, vibration, microwave, temperature, ultrasonic, pH/chemical, and pressure, were systematically examined. Each method was analysed in terms of its underlying principles, key influencing factors, and potential applications in modern construction practices. These techniques offer innovative ways to dynamically adjust the rheological properties of fresh cementitious materials.

Despite the promising developments in this emerging field, several challenges remain, including scalability and cost-effectiveness. Future research should focus on optimizing these control strategies and developing responsive smart systems that can adapt to real-time construction demands. Additionally, further investigations into their effects on hydration and microstructure evolution are needed.

References

1. Jiao, D., Shi, C., Yuan, Q., et al.: Effect of constituents on rheological properties of fresh concrete—a review. Cem. Concr. Compos. **83**, 146–159 (2017)
2. Li, F., Shen, W., Yuan, Q., et al.: An overview on the effect of pumping on concrete properties. Cem. Concr. Compos. **129** (2022)
3. Feys, D., De Schutter, G., Verhoeven, R., et al.: Similarities and differences of pumping conventional and self-compacting concrete. In: Design, Production and Placement of Self-Consolidating Concrete, pp. 153–162. Springer (2010)
4. Kaplan, D.: Pumping of concretes, Ph.D. dissertation (in French). Laboratoire Central des Ponts et Chausées, Paris (2001)
5. Choi, M.S., Kim, Y.J., Kwon, S.H.: Prediction on pipe flow of pumped concrete based on shear-induced particle migration. Cem. Concr. Res. **52**, 216–224 (2013)
6. De Schutter, G., Feys, D.: Pumping of fresh concrete: insights and challenges. RILEM Tech. Lett. **1**, 76–80 (2016)
7. Secrieru, E., Cotardo, D., Mechtcherine, V., et al.: Changes in concrete properties during pumping and formation of lubricating material under pressure. Cem. Concr. Res. **108**, 129–139 (2018)
8. De Schutter, G.: Thixotropic effects during large-scale concrete pump tests on site. In: 71st RILEM Annual Week & ICACMS 2017, Chennai, India (2017)
9. Roussel, N., Cussigh, F.: Distinct-layer casting of SCC: The mechanical consequences of thixotropy. Cem. Concr. Res. **38**(5), 624–632 (2008)
10. Ovarlez, G., Roussel, N.: A physical model for the prediction of lateral stress exerted by self-compacting concrete on formwork. Mater. Struct. **39**(2), 269–279 (2006)
11. Roussel, N.: A thixotropy model for fresh fluid concretes: theory, validation and applications. Cem. Concr. Res. **36**(10), 1797–1806 (2006)
12. Jiao, D., Shi, C., Yuan, Q., et al.: Mixture design of concrete using simplex centroid design method. Cem. Concr. Compos. **89**, 76–88 (2018)
13. Ashish, D.K., Verma, S.K.: An overview on mixture design of self-compacting concrete. Struct. Concr. **20**(1), 371–395 (2019)
14. Han, F., Li, Y., Jiao, D.: Understanding the rheology and hydration behaviour of cement paste with nickel slag. J. Build. Eng. **73**, 106724 (2023)
15. Ma, Y., Jiao, D., Sha, S., et al.: Effect of anchoring groups of polycarboxylate ether superplasticizer on the adsorption and dispersion of cement paste containing montmorillonite. Cem. Concr. Compos. **134** (2022)
16. Uchikawa, H., Sawaki, D., Hanehara, S.: Influence of kind and added timing of organic admixture on the composition, structure and property of fresh cement paste. Cem. Concr. Res. **25**(2), 353–364 (1995)
17. Benaicha, M., Roguiez, X., Jalbaud, O., et al.: Influence of silica fume and viscosity modifying agent on the mechanical and rheological behaviour of self compacting concrete. Constr. Build. Mater. **84**, 103–110 (2015)
18. De Schutter, G., Lesage, K.: Active control of properties of concrete: a (p)review. Mater. Struct. **51**(5), 123 (2018)

19. De Schutter, G., Lesage, K.: Active Rheology Control of Cementitious Materials. CRC Press (2022)
20. De Schutter, G., Abidi, I., Korda, E., et al.: Active control of properties of fresh and hardening concrete. RILEM Tech. Lett. **8**, 198–203 (2023)
21. Banfill, P.: The rheology of fresh cement and concrete-rheology review. Br. Soc. Rheol. **61**, 130 (2006)
22. Guo, X., Jiao, D.: Rheology as a versatile tool in concrete technology: a mini-review. Adv. Manuf. **2**, 0006 (2024)
23. Banfill, P.: The rheology of cement paste: progress since 1973. In: Properties of Fresh Concrete: Proceedings of the International RILEM Colloquium, pp. 3–9. CRC Press (1990)
24. Ma, S., Qian, Y., Kawashima, S.: Experimental and modeling study on the non-linear structural build-up of fresh cement pastes incorporating viscosity modifying admixtures. Cem. Concr. Res. **108**, 1–9 (2018)
25. Roussel, N., Ovarlez, G., Garrault, S., et al.: The origins of thixotropy of fresh cement pastes. Cem. Concr. Res. **42**(1), 148–157 (2012)
26. Jiao, D., Shi, C., Yuan, Q.: Time-dependent rheological behaviour of cementitious paste under continuous shear mixing. Constr. Build. Mater. **226**, 591–600 (2019)
27. Barnes, H.A.: Thixotropy—a review. J. Nonnewton. Fluid Mech. **70**(1–2), 1–33 (1997)
28. Jiao, D., De Schryver, R., Shi, C., et al.: Thixotropic structural build-up of cement-based materials: a state-of-the-art review. Cem. Concr. Compos. **122**, 104152 (2021)
29. Møller, P.C., Mewis, J., Bonn, D.: Yield stress and thixotropy: on the difficulty of measuring yield stresses in practice. Soft Matter **2**(4), 274–283 (2006)
30. Jiao, D., De Schutter, G.: Insights into the viscoelastic properties of cement paste based on SAOS technique. Constr. Build. Mater. **357**, 129320 (2022)
31. Qian, Y.: Effect of polycarboxylate ether (PCE) superplasticizer on thixotropic structural build-up of fresh cement pastes over time. Constr. Build. Mater. **291**, 123241 (2021)
32. Qian, Y., Kawashima, S.: Distinguishing dynamic and static yield stress of fresh cement mortars through thixotropy. Cement Concr. Compos. **86**, 288–296 (2018)
33. Cheng, X., McCoy, J.H., Israelachvili, J.N., et al.: Imaging the microscopic structure of shear thinning and thickening colloidal suspensions. Science **333**(6047), 1276–1279 (2011)
34. Lu, S., Mei, Q., Chen, J., et al.: Cryo-TEM and rheological study on shear-thickening wormlike micelles of zwitterionic/anionic (AHSB/SDS) surfactants. J. Colloid Interface Sci. **608**, 513–524 (2022)
35. Vázquez-Quesada, A., Mahmud, A., Dai, S., et al.: Investigating the causes of shear-thinning in non-colloidal suspensions: experiments and simulations. J. Nonnewton. Fluid Mech. **248**, 1–7 (2017)
36. Yuan, Q., Shi, C., Jiao, D.: Rheology of Fresh Cement-Based Materials: Fundamentals, Measurements, and Applications. CRC Press (2022)
37. Huang, F., Li, H., Yi, Z., et al.: The rheological properties of self-compacting concrete containing superplasticizer and air-entraining agent. Constr. Build. Mater. **166**, 833–838 (2018)
38. Chia, K., Zhang, M.: Effect of chemical admixtures on rheological parameters and stability of fresh lightweight aggregate concrete. Mag. Concr. Res. **56**(8), 465–473 (2004)
39. Yahia, A.: Shear-thickening behaviour of high-performance cement grouts—influencing mix-design parameters. Cem. Concr. Res. **41**(3), 230–235 (2011)
40. Zhu, J., Xie, W., Li, Z., et al.: An approach to describe the shear-thickening viscosity of cement paste incorporating microfines of manufactured sand. Constr. Build. Mater. **340**, 127743 (2022)
41. Shen, W., Cui, H., Jiao, D.: Experimental observations on factors influencing shear-thickening characteristics of cement-based materials. J. Build. Eng. **86** (2024)
42. Jiao, D., Shi, C., Yuan, Q.: Influences of shear-mixing rate and fly ash on rheological behaviour of cement pastes under continuous mixing. Constr. Build. Mater. **188**, 170–177 (2018)
43. Jiao, D., Shi, C., Yuan, Q., et al.: Effects of rotational shearing on rheological behaviour of fresh mortar with short glass fiber. Constr. Build. Mater. **203**, 314–321 (2019)

44. Ma, K., Feng, J., Long, G., et al.: Effects of mineral admixtures on shear thickening of cement paste. Constr. Build. Mater. **126**, 609–616 (2016)
45. Rabinow, J.: The magnetic fluid clutch. Electr. Eng. **67**(12), 1167–1167 (1948)
46. Park, J.H., Chin, B.D., Park, O.O.: Rheological properties and stabilization of magnetorheological fluids in a water-in-oil emulsion. J. Colloid Interface Sci. **240**(1), 349–354 (2001)
47. Vinod, S., John, R., Philip, J.: Magnetorheological properties of sodium sulphonate capped electrolytic iron based MR fluid: a comparison with CI based MR fluid. Smart Mater. Struct. **26**(2) (2017)
48. Wereley, N.M., Chaudhuri, A., Yoo, J.H., et al.: Bidisperse magnetorheological fluids using Fe particles at nanometer and micron scale. J. Intell. Mater. Syst. Struct. **17**(5), 393–401 (2006)
49. Genç, S., Phulé, P.P.: Rheological properties of magnetorheological fluids. Smart Mater. Struct. **11**(1), 140–146 (2002)
50. Ahamed, R., Choi, S.-B., Ferdaus, M.M.: A state of art on magneto-rheological materials and their potential applications. J. Intell. Mater. Syst. Struct. **29**(10), 2051–2095 (2018)
51. Bica, I., Liu, Y.D., Choi, H.J.: Physical characteristics of magnetorheological suspensions and their applications. J. Ind. Eng. Chem. **19**(2), 394–406 (2013)
52. Schultz, M.A., Struble, L.J.: Use of oscillatory shear to study flow behaviour of fresh cement paste. Cem. Concr. Res. **23**(2), 273–282 (1993)
53. Tattersall, G.H., Banfill, P.F.: The rheology of fresh concrete (1983)
54. Jiao, D.: Magneto-rheology control of cementitious materials: principles, current status, and opportunities. RILEM Tech. Lett. **10**, 44–50 (2025)
55. Nair, S.D., Ferron, R.D.: Set-on-demand concrete. Cem. Concr. Res. **57**, 13–27 (2014)
56. Jiao, D., Lesage, K., Yardimci, M.Y., et al.: Rheological behaviour of cement paste with nano-Fe_3O_4 under magnetic field: magneto-rheological responses and conceptual calculations. Cem. Concr. Compos. **120**, 104035 (2021)
57. Jiao, D., Lesage, K., Yardimci, M.Y., et al.: Structural evolution of cement paste with nano-Fe_3O_4 under magnetic field—effect of concentration and particle size of nano-Fe_3O_4. Cem. Concr. Compos. **120**, 104036 (2021)
58. Jiao, D., Lesage, K., Yardimci, M.Y., et al.: Quantitative assessment of the influence of external magnetic field on clustering of nano-Fe_3O_4 particles in cementitious paste. Cem. Concr. Res. **142**, 106345 (2021)
59. Jiao, D., Shi, C., De Schutter, G.: Estimation of magnetic force between micrometer-sized fly-ash particles in cementitious suspensions. J. Mater. Civ. Eng. **35**(2), 04022421 (2023)
60. Jiao, D., Yardimci, M.Y., De Schryver, R., et al.: Effect of solid volume fraction on the magnetorheological response of nano-Fe_3O_4 incorporated cementitious paste. J. Mater. Civ. Eng. **34**(11), 04022280 (2022)
61. Jiao, D., Shi, C., De Schutter, G.: Magneto-responsive structural build-up of highly flowable cementitious paste in the presence of PCE superplasticizer. Constr. Build. Mater. **327**, 126925 (2022)
62. Jiao, D., El Cheikh, K., Lesage, K., et al.: Structural build-up of cementitious paste under external magnetic fields, pp. 36–42. Springer, Rheol. Process. Constr. Mater. (2020)
63. Jiao, D., El Cheikh, K., Shi, C., et al.: Structural build-up of cementitious paste with nano-Fe_3O_4 under time-varying magnetic fields. Cem. Concr. Res. **124**, 105857 (2019)
64. Krezinski, C., Panchmatia, P., Genedy, M., et al.: Magnetorheological cements for plug setting control in oil and gas wells. J. Petrol. Sci. Eng. **200**, 108257 (2021)
65. Kim, J., Ferron, R.D.: Effect of air-entraining agent on rheology of cement-based magnetorheological pastes. ACI Mater. J. **121**(5), 3–12 (2024)
66. Chibulu, C., Yardimci, M.Y., Jiao, D., et al.: Active stiffening control by magnetically induced blocking in confined flow of fly ash pastes. Constr. Build. Mater. **313**, 125485 (2021)
67. Chibulu, C., Jiao, D., Yardimci, M., et al.: Magneto-rheology control of cement paste containing Fe_3O_4 nanoparticles in view of reducing or preventing formwork leakage. Cem. Concr. Compos. **142**, 105176 (2023)

68. Nair, S.D., Ferron, R.D.: Real time control of fresh cement paste stiffening: smart cement-based materials via a magnetorheological approach. Rheol. Acta **55**(7), 571–579 (2016)
69. Kanagasuntharam, S., Ramakrishnan, S., Muthukrishnan, S., et al.: Effect of Magnetorheological additives on the buildability of 3D concrete printing. J. Build. Eng. **74** (2023)
70. Zhang, Y., Lesage, K., Zhang, Y., et al.: A comparison of magneto-responsive particles and testing protocols for active rheology control of cementitious materials. Cem. Concr. Compos. **146**, 105390 (2024)
71. Jiao, D., Yardimci, M.Y., Lesage, K., et al.: Active rheology control of cementitious materials with responsive mineral particles. Active Rheol. Control Cementitious Mater. 77–136 (2023)
72. Zhang, Y., Tao, Y., Sun, Y., et al.: Active control of thixotropy of magneto-responsive cementitious materials with the intervention of time-varying magnetic fields. Cem. Concr. Res. **181**, 107520 (2024)
73. Jiao, D., Lesage, K., Yardimci, M.Y., et al.: Rheological properties of cement paste with nano-Fe_3O_4 under magnetic field: flow curve and nanoparticle agglomeration. Materials **13**(22), 5164 (2020)
74. Jiao, D., Lesage, K., Yardimci, M.Y., et al.: Flow behaviour of cementitious-like suspension with nano-Fe_3O_4 particles under external magnetic field. Mater. Struct. **54**(6), 209 (2021)
75. Jiao, D., Hu, S., De Schutter, G.: How does ambient temperature affect the magneto-rheological response of cementitious paste? Constr. Build. Mater. **490**, 142543 (2025)
76. Jiao, D., De Schutter, G.: Enhancing magneto-rheology control of cement paste through synergistic effect of nano-Fe_3O_4 particles and fly ash. Mater. Lett. **367**, 136630 (2024)
77. Deshmukh, A.S., Heintzkill, R.T., Huerta, R.A., et al.: Rheological response of magnetorheological cementitious inks tuned for active control in digital construction. ACI Mater. J. **118**(6), 263–274 (2021)
78. Jiao, D., Lesage, K., Yardimci, M.Y., et al.: Possibilities of fly ash as responsive additive in magneto-rheology control of cementitious materials. Constr. Build. Mater. **296**, 123656 (2021)
79. Yardimci, M.Y., Jiao, D., De Schutter, G.: Magnetorheological properties of fayalite slag incorporated cement mixtures. In: RILEM Spring Convention and Conference. Springer Nature Switzerland Cham, pp. 238–246 (2024)
80. Heik, P., Ehle, F., Echt, T. et al.: Active rheology control of cementitious materials: Comparing residual materials with potentially magnetic particles. In: Posani, M., Singh, S. (eds) RILEM Youth Symposium 2025 – RYS2025. RYS 2025. RILEM Bookseries, vol 67. Springer, Cham. (2026)
81. Heik, P., Ehle, F., Echt, T. et al.: Magneto-responsive residual materials for active rheology control in cement pastes. Constr. Build. Mater. **533**, 146828 (2026)
82. Jiao, D., Shi, C., De Schutter, G.: Magneto-rheology control in 3D concrete printing: a rheological attempt. Mater. Lett. **309**, 131374 (2021)
83. Davies, B., Mottram, R., Harris, I.: Recent developments in the sintering of NdFeB. Mater. Chem. Phys. **67**(1–3), 272–281 (2001)
84. Zhang, Y., Zhang, Y., Tao, Y., et al.: Active rheology control of cementitious materials containing hard magnetic particles: sustained response after magnetic intervention. Cem. Concr. Compos. 106024 (2025)
85. Zhang, Y., Tao, Y., Zhang, Y., et al.: Upscaling active rheology control to cement mortar with the intervention of an inline magnetic field. Constr. Build. Mater. **425**, 135853 (2024)
86. Zhang, Y., Tao, Y., Godinho, J.R., et al.: Active rheology control of cementitious materials: new insights from magneto-responsive aggregates. Cem. Concr. Res. **189**, 107707 (2025)
87. Guo, X., Hu, S., Jiao, D.: Flowability responses of cement paste containing steel fiber to external magnetic field. J. Build. Eng. **103**, 112220 (2025)
88. Hu, F., Wu, H., Jiao, D.: Effect of magneto-rheological response on early hydration and microstructure of cement paste. Constr. Build. Mater. **474**, 141161 (2025)
89. Liang, Y., Hu, F., Wu, H., et al.: Understanding the role of magneto-rheological control in modifying the mechanical characteristics of cement pastes. Constr. Build. Mater. **499**, 144091 (2025)

90. Block, H., Kelly, J.: Electro-rheology. J. Phys. D Appl. Phys. **21**(12), 1661 (1988)
91. Halsey, T.C.: Electrorheological fluids. Science **258**(5083), 761–766 (1992)
92. Otsubo, Y., Edamura, K.: Static yield stress of electrorheological fluids. J. Colloid Interface Sci. **172**(2), 530–535 (1995)
93. De Schryver, R., Jiao, D., Yardimci, M.Y., et al.: A conceptual model for numerical simulation of active rheology control validated for cementitious pumping. Cem. Concr. Res. **179** (2024)
94. Choi, M.S., Kim, Y.S., Kim, J.H., et al.: Effects of an externally imposed electromagnetic field on the formation of a lubrication layer in concrete pumping. Constr. Build. Mater. **61**, 18–23 (2014)
95. Goudjil, N., Vanhove, Y., Djelal, C., et al.: Electro-osmosis applied for formwork removal of concrete. J. Adv. Concr. Technol. **10**(9), 301–312 (2012)
96. Goudjil, N., Djelal, C., Vanhove, Y., et al.: Impact of temperature on the demoulding of concrete elements with a polarization process. Constr. Build. Mater. **54**, 402–412 (2014)
97. Muhammad, M.: Concrete superplasticizers: from fundamental studies to the development of smart superplasticizers for rheology control. Ghent University (2022)
98. De Schutter, G., Ezzat, M., Lesage, K., et al.: Responsive superplasticizers for active rheology control of cementitious materials. Cem. Concr. Res. **165**, 107084 (2023)
99. Ezzat, M., Lesage, K., Hoogenboom, R., et al.: Smart superplasticizers based on redox-responsive polymers for rheology control of cementitious materials. Cem. Concr. Res. **173**, 107291 (2023)
100. Cai, Y., Liu, Q.-F., Yu, L., et al.: An experimental and numerical investigation of coarse aggregate settlement in fresh concrete under vibration. Cem. Concr. Compos. **122**, 104153 (2021)
101. Tattersall, G.H., Baker, P.H.: The effect of vibration on the rheological properties of fresh concrete. Mag. Concr. Res. **40**(143), 79–89 (1988)
102. Tattersall, G., Baker, P.: An investigation on the effect of vibration on the workability of fresh concrete using a vertical pipe apparatus. Mag. Concr. Res. **41**(146), 3–9 (1989)
103. Li, Z., Cao, G.: Rheological behaviours and model of fresh concrete in vibrated state. Cem. Concr. Res. **120**, 217–226 (2019)
104. Sanjayan, J.G., Jayathilakage, R., Rajeev, P.: Vibration induced active rheology control for 3D concrete printing. Cem. Concr. Res. **140**, 106293 (2021)
105. Korson, L., Drost-Hansen, W., Millero, F.J.: Viscosity of water at various temperatures. J. Phys. Chem. **73**(1), 34–39 (1969)
106. Wang, M., Zhu, Z., Liu, R., et al.: Influence of extreme high-temperature environment and hydration time on the rheology of cement slurry. Constr. Build. Mater. **295**, 123684 (2021)
107. Zhang, Y., Jiang, Z., Zhu, Y., et al.: Effects of redispersible polymer powders on the structural build-up of 3D printing cement paste with and without hydroxypropyl methylcellulose. Constr. Build. Mater. **267**, 120551 (2021)
108. Jang, S.P., Choi, S.U.: Role of Brownian motion in the enhanced thermal conductivity of nanofluids. Appl. Phys. Lett. **84**(21), 4316–4318 (2004)
109. Zhang, Y., Ren, Q., Dai, X., et al.: A potential active rheology control approach for 3D printable cement-based materials: coupling of temperature and viscosity modifiers. Cem. Concr. Compos. **149**, 105496 (2024)
110. Muthukrishnan, S., Ramakrishnan, S., Sanjayan, J.: Effect of microwave heating on interlayer bonding and buildability of geopolymer 3D concrete printing. Constr. Build. Mater. **265**, 120786 (2020)
111. Qu, Z., Yu, Q., Ong, G.P., et al.: 3D printing concrete containing thermal responsive gelatin: towards cold environment applications. Cem. Concr. Compos. **140**, 105029 (2023)
112. Christ, J., Perrot, A., Ottosen, L.M., et al.: Rheological characterization of temperature-sensitive biopolymer-bound 3D printing concrete. Constr. Build. Mater. **411**, 134337 (2024)
113. Ramakrishnan, S., Kanagasuntharam, S., Sanjayan, J.: In-line activation of cementitious materials for 3D concrete printing. Cem. Concr. Compos. 104598 (2022)

114. Shao, L., Feng, P., Zuo, W., et al.: A novel method for improving the printability of cement-based materials: controlling the releasing of capsules containing chemical admixtures. Cem. Concr. Compos. **128**, 104456 (2022)
115. Xuequan, W., Jianbgo, D., Mingshu, T.: Microwave curing technique in concrete manufacture. Cem. Concr. Res. **17**(2), 205–210 (1987)
116. Leung, C.K., Pheeraphan, T.: Microwave curing of Portland cement concrete: experimental results and feasibility for practical applications. Constr. Build. Mater. **9**(2), 67–73 (1995)
117. Aslam, R., Alam, M.S., Kaur, J., et al.: Understanding the effects of ultrasound processing on texture and rheological properties of food. J. Texture Stud. **53**(6), 775–799 (2022)
118. Kabbani, D., Sepulcre, F., Wedekind, J.: Ultrasound-assisted liquefaction of rosemary honey: influence on rheology and crystal content. J. Food Eng. **107**(2), 173–178 (2011)
119. Amiri, A., Sharifian, P., Soltanizadeh, N.: Application of ultrasound treatment for improving the physicochemical, functional and rheological properties of myofibrillar proteins. Int. J. Biol. Macromol. **111**, 139–147 (2018)
120. Martínez-Velasco, A., Lobato-Calleros, C., Hernández-Rodríguez, B.E., et al.: High intensity ultrasound treatment of faba bean (Vicia faba L.) protein: effect on surface properties, foaming ability and structural changes. Ultrason. Sonochem. **44**, 97–105 (2018)
121. Sauter, C., Emin, M., Schuchmann, H., et al.: Influence of hydrostatic pressure and sound amplitude on the ultrasound induced dispersion and de-agglomeration of nanoparticles. Ultrason. Sonochem. **15**(4), 517–523 (2008)
122. Xu, Z., Ji, Y., Ma, Z., et al.: Strengthening mechanism of ultrasonic action on mechanical properties of cement-based materials. Constr. Build. Mater. **362**, 129788 (2023)
123. Wang, X., Huang, J., Dai, S., et al.: Effect of silica fume particle dispersion and distribution on the performance of cementitious materials: a theoretical analysis of optimal sonication treatment time. Constr. Build. Mater. **212**, 549–560 (2019)
124. Silvestro, L., Ruviaro, A., Lima, G., et al.: Influence of ultrasonication of functionalized carbon nanotubes on the rheology, hydration, and compressive strength of portland cement pastes. Materials **14**(18), 5248 (2021)
125. Silvestro, L., Ruviaro, A.S., Lima, G.T.D.S., et al.: Sonicating polycarboxylate-based superplasticizer for application in cementitious matrix. Revista IBRACON de Estruturas e Materiais **16**(2), e16205 (2022)
126. Zou, B., Chen, S.J., Korayem, A.H., et al.: Effect of ultrasonication energy on engineering properties of carbon nanotube reinforced cement pastes. Carbon **85**, 212–220 (2015)
127. Liu, J., Fu, J., Yang, Y., et al.: Study on dispersion, mechanical and microstructure properties of cement paste incorporating graphene sheets. Constr. Build. Mater. **199**, 1–11 (2019)
128. De La Rosa, Á., Ruiz, G., Husillos-Rodríguez, N., et al.: Effect of cellulose nanofibers on the rheological and mechanical properties of cement pastes with metakaolin and natural volcanic pozzolan. Constr. Build. Mater. **452**, 138823 (2024)
129. Puertas, F., Gismera, S., Lanzón, M., et al.: Rheology of cementitious materials: alkali-activated materials or geopolymers. In: MATEC Web of Conferences, EDP Sciences, p. 01002 (2018)
130. Zhang, Q.-L., Li, Y.-T., Chen, Q.-S., et al.: Effects of temperatures and pH values on rheological properties of cemented paste backfill. J. Cent. S. Univ. **28**(6), 1707–1723 (2021)
131. Kanagasuntharam, S., Ramakrishnan, S., Sanjayan, J.: Encapsulation of sodium silicate to attain on demand buildability enhancement in concrete 3D printing. J. Build. Eng. **94**, 109912 (2024)
132. Muthukrishnan, S., Ramakrishnan, S., Sanjayan, J.: Set on demand geopolymer using print head mixing for 3D concrete printing. Cem. Concr. Compos. **128** (2022)
133. Ouchi, M., Sakue, J.: Self-compactability of fresh concrete in terms of dispersion and coagulation of particles of cement subject to pumping. In: Proceedings of the 3rd North-American Conference on the Design and Use of Self-Consolidating Concrete, Chicago (2008)
134. Feys, D., Khayat, K.H., Khatib, R.: How do concrete rheology, tribology, flow rate and pipe radius influence pumping pressure? Cem. Concr. Compos. **66**, 38–46 (2016)

135. Hover, K.C.: Some recent problems with air-entrained concrete. Cem. Concr. Aggreg. **11**(1), 67–72 (1989)
136. Pleau, R., Pigeon, M., Lamontagne, A., et al.: Influence of pumping on characteristics of air-void system of high-performance concrete. Transp. Res. Rec. 30–36 (1995)
137. Feys, D., Roussel, N., Verhoeven, R., et al.: Influence of air content on the steady state rheological properties of fresh self-compacting concrete, without air entraining agents. In: 2nd International symposium on Design, Performance and Use of Self-Consolidating Concrete (SCC2009), RILEM Publications, pp. 287–295 (2009)
138. Kuder, K.G., Shah, S.P.: Rheology of extruded cement-based materials. ACI Mater. J. **104**(3), 283 (2007)
139. Farris, R.F.: Effects of temperature and pressure on rheological properties of cement slurries. Trans. AIME **142**(01), 117–130 (1941)
140. Yim, H.J., Kim, J.H., Kwon, S.H.: Effect of admixtures on the yield stresses of cement pastes under high hydrostatic pressures. Materials **9**(3), 147 (2016)
141. Kim, J.H., Kwon, S.H., Kawashima, S., et al.: Rheology of cement paste under high pressure. Cem. Concr. Compos. **77**, 60–67 (2017)
142. Liu, K., Cheng, X., Gao, X., et al.: Effect of the hydration rate and microstructure of Portland cement slurry on hydrostatic pressure transfer. Powder Technol. **352**, 251–261 (2019)
143. Zhang, Q., Shu, X., Yang, Y., et al.: Effect of high pressure on the time-dependent rheological properties of cement paste. J. Mater. Civ. Eng. **35**(10), 04023359 (2023)

Active Control of Setting of Cementitious Materials

Shravan Muthukrishnan, Luise Göbel, Ojas Arun Chaudhari, Paula Heik, Yi Zhang, and Yaxin Tao

Abstract The setting rate of concrete is a key factor influencing the productivity of construction processes, including the determination of demoulding time in conventional in-situ casting and the transfer timing of 3D-printed modules from the print bed. To achieve consistent and enhanced productivity in dynamic construction environments active control over the setting behaviour of cementitious materials is essential. This chapter provides an overview of various strategies developed to achieve active setting control in Portland cement-based systems and alkali-activated binders. The methods discussed encompass chemical approaches (e.g., accelerator admixtures), thermal techniques (e.g., microwave-assisted setting), and mechanical interventions (e.g., ultrasonic-induced setting). Furthermore, the chapter delineates the distinction between active rheology control (see chapter "Active Control of Rheology of Cementitious Materials") and active setting control, underlining how both aspects complement each other in fulfilling the performance demands of modern construction processes. The discussion concludes with future perspectives and research directions derived from the collective expertise of the contributing authors.

S. Muthukrishnan (✉)
Cluster of Excellence CARE, TU Dresden and RWTH Aachen, Dresden, Germany
e-mail: shravan.muthukrishnan@tu-dresden.de

Institute of Construction Materials, TUD Dresden University of Technology, Dresden, Germany

L. Göbel · P. Heik
F. A. Finger-Institute for Building Materials Science, Chair of Mechanics of Engineering Materials, Bauhaus-Universität Weimar, Weimar, Germany

O. A. Chaudhari
RISE - Brinellgatan, Borås, Sweden

Y. Zhang
Department of Structural Engineering and Building Materials, Magnel-Vandepitte Laboratory, Ghent University, Ghent, Belgium

Y. Tao
Department of Geotechnical Engineering, Tongji University, Shanghai, China

G. De Schutter and J. Sanjayan (eds.), *Active Control of Properties of Fresh and Hardening Cementitious Materials*, RILEM State-of-the-Art Reports 50,
https://doi.org/10.1007/978-3-032-30340-0_4

Keywords 3D concrete printing · Set-on-demand concrete · Low carbon concrete · Digital construction · Buildability

1 Introduction

Shortly after water or activators are added to cementitious systems, the mix begins to gain strength due to a combination of flocculation (a physical effect) and chemical reactions. Figure 1 illustrates the strength development of concrete over time after wet mixing. In the early stages, the strengthening process is primarily governed by flocculation. However, at a certain point, the chemical reactions begin to dominate, surpassing the effects of physical flocculation [1]. The active control of strength development during the earlier phase, dominated by rheological properties, is discussed in chapter "Active Control of Rheology of Cementitious Materials": Active Control on Rheological Properties. This chapter focuses on the active control of strength development (referred to as setting hereafter) beyond this transition point. The setting rate of concrete is a critical factor influencing the productivity of construction techniques. For instance, in traditional casting methods, an accelerated setting rate can reduce demoulding time, while in 3D concrete printing (3DCP), it determines the rate at which layers can be deposited one over the other, which is known as buildability. However, accelerated setting is not always advantageous and must be carefully regulated depending on the specific application. For example, in 3DCP, excessive setting can significantly compromise interlayer bond adhesion.

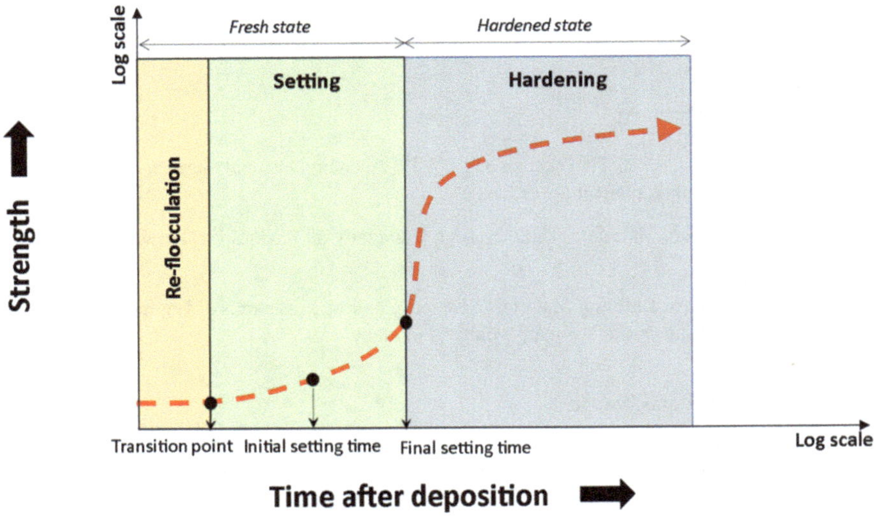

Fig. 1 Strength development of concrete with time after wet mixing. The figure was reproduced with permission from Ref. [1]

Additionally, in most construction techniques, concrete is required to exhibit contradictory setting properties at different stages of the process. For example, in traditional casting methods, the setting rate should be slow during pumping to ensure a longer open time and adequate workability. Conversely, during mould filling (casting), a faster setting rate is desirable to minimize the formwork pressure. Similarly, in modern technologies like 3D concrete printing (3DCP), the concrete must set slowly during pumping. However, once deposited, the layers need to set quickly to enhance buildability. Active control of the concrete setting process enables achieving these contradictory fresh properties, which would otherwise be challenging to accomplish. This chapter discusses various methods investigated in the literature for achieving active control of the setting in both Portland cement concrete and alkali-activated materials, applicable to traditional and modern construction techniques.

2 Chemical Induced Setting

2.1 Accelerators

Accelerators primarily function by increasing the concentration of ions, such as Ca^{2+} and SO_4^{2-}, in the cement paste [2]. This effect is achieved either by directly introducing salts containing such ions or by promoting the dissolution of cement particles [3]. This elevation in ionic concentration significantly enhances the reaction rates of key cement phases, including tricalcium aluminate (C_3A), tricalcium silicate (C_3S), and dicalcium silicate (C_2S), thereby shortening the setting time of cement-based materials. The active setting control by accelerators primarily works by regulating the timing of their introduction and the way they interact with cement particles. Liquid accelerators are commonly introduced at the spray nozzle in wet-mixed shotcrete applications to achieve quick setting of the cement-based materials, providing immediate structural support during construction (Fig. 2) [4]. Powder-based accelerators can be blended with raw materials in dry-mix shotcrete applications, where they are activated upon water addition at the spray nozzle, accelerating cement hydration [5]. In cast concrete applications, accelerators are sometimes added and blended with the mixture shortly prior to casting to obtain required setting rate. Based on the same principles, recently, accelerators have been used for 3D concrete printing to achieve a good buildability after material deposition [6, 7]. By incorporating a secondary mixing process at the nozzle, including dynamic and static mixing, where wet mixtures are combined with accelerators, the setting rate is significantly enhanced. To eliminate the need for secondary mixing, alternative approaches have been proposed that utilize external fields, such as temperature and ultrasonic waves. These methods aim to trigger the release of accelerators encapsulated in capsules or to activate the efficacy of pre-mixed accelerators into the mixtures through external field. More details on this are explained in Sect. 5.

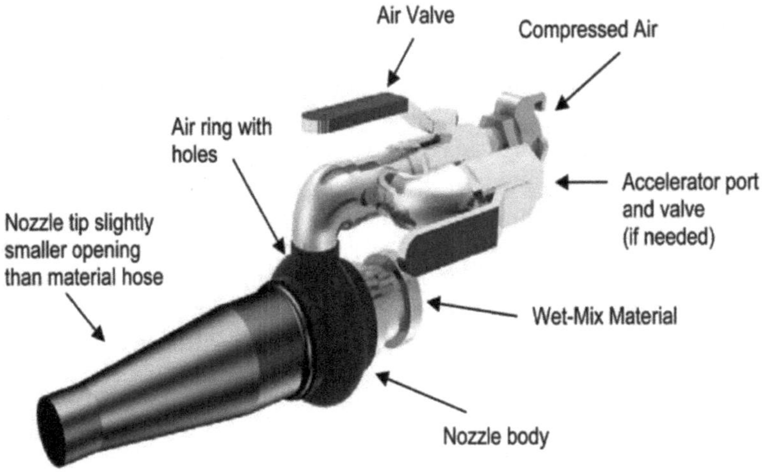

Fig. 2 A typical nozzle used to attain active setting control in wet-mix shotcrete application. The figure is reproduced with permission from Ref. [9]

Active control of concrete setting by injecting accelerators at the print head is widely adopted, owing to the accelerators' availability, stability, and ease of application. However, the rapid setting induced by these accelerators can adversely affect the long-term properties of the concrete. Furthermore, these admixtures are highly sensitive to the composition of the base mix prior to activation at the print head [8]. This sensitivity can be mitigated by employing rapid-setting cement suspensions, as discussed in the next section.

2.2 Rapid Setting Cement Suspension

Rapid setting cement suspensions are usually made by rapid setting cement, such as calcium aluminate cement (CAC), calcium sulfoaluminate cement (CSA), magnesium phosphate cement (MPC) [10]. These suspensions are primarily utilized in emergency repair engineering due to their ability to quickly set. Unlike Ordinary Portland Cement (OPC), these cements feature distinct mineral compositions with aluminate phases, such as ye'elimite in CSA or calcium aluminate phases in CAC and often exhibit faster hydration processes. Rapid setting cement can be considered as accelerators to blend with OPC, therefore shortening the setting time of blended cement-based materials [11]. These cement suspensions set very rapidly, which can pose challenges during on-site mixing.

In the case of long-distance transport and pumping, active setting control for these materials is essential, which typically involves both retarding and accelerating stages. During transportation and pumping, retarders are employed to delay hydration and maintain workability. Just before casting, accelerators or admixtures capable of counteracting the retardation effects are added to initiate rapid hydration, ensuring fast setting. Tao et al. [12] adopted borax ($Na_2B_4O_7\cdot 10H_2O$) to delay hydration of CSA cement-based mixture during pumping, while using the alkali substrates in OPC or additional calcium hydroxide (CH) to eliminate the retardation effect of borax after pumping. The retardation mechanism involves the formation of a complex between $B_4O_7^{2-}$ ions and Ca^{2+} ions, which leads to the creation of a calcium borate layer that adsorbs onto cement particles. This layer inhibits the hydration process, effectively delaying the setting and hardening of the cement-based mixture [13]. The addition of CH can react with borax, and it provides extra Ca^{2+} and OH^-, which compensate for the retardation effects and accelerates hydration reactions. Certain components, like clay, can adsorb retarders and water reducers, hence counteracting the retarding effect of cement, therefore can be used as an alternative to CH [14].

Most retarders and accelerators rely on the presence of calcium ions, which are primarily supplied by OPC in the mixture. However, OPC is known for its significant environmental impact due to its carbon-intensive production process. To mitigate this, researchers have investigated alternative materials to partially or fully replace OPC in construction applications. The use of low carbon low clinker cements can change the chemistry of the binder systems. Fully replacing OPC (Alkali-activated materials (AAMs)) in concrete makes conventional retarders and accelerators ineffective for achieving active setting control. The next section outlines the strategies employed in AAMs to achieve active setting control.

2.3 Alkali Activated Materials

AAMs represent a promising solution, where the OPC is entirely substituted with precursors from different resources, such as fly ash, calcined clay, and slag. These precursors require an alkaline environment for activation, which is provided by OPC in cases of partial substitution. However, in the case of full substitution, additional alkaline activators are necessary—hence the name "alkali-activated" materials (AAMs). The AAM slurry, comprising of SCMs (hereafter referred to as "precursors" in the context of AAMs terminology) and water does not harden due to the insufficient alkaline environment for activation. Building on this understanding, Muthukrishnan et al. [15, 16] explored an active setting control strategy, in which AAM slurry with an extended open time is pumped into an in-line mixer. There, it is combined with an alkaline activator to regulate the setting rate of concrete, as required by the construction process. It was found the setting rate of AAM can be controlled by varying the activator dosage and silicate modulus. According to [15], for a constant silicate modulus (SiO_2/Na_2O), increasing the activator dosage initially accelerates the setting rate of concrete. However, beyond a certain point, further increase in the activator dosage leads to a decrease in the setting rate. This can be attributed to the excessive alkalinity of the pore solution, which inhibits the formation of reaction products. While alkalinity is essential for dissolving reactive components, excessive levels compromise the stability of the reaction products. Therefore, achieving the optimal activator dosage is critical to balancing dissolution and stability, ensuring the desired setting behaviour.

The active control over the setting rate of alkali-activated materials (AAMs) can also be achieved by adjusting the silicate modulus. This was done by varying the dosages of sodium silicate and sodium hydroxide solutions fed into the inline mixer. Study found that increasing the silicate modulus from 0.34 to 1.72 reduced the initial static yield stress. However, further increases in the silicate modulus beyond 1.72 resulted in higher initial static yield stress [17]. This behaviour was attributed to the dual effects of silicate anions. At lower concentrations, silicate anions enhanced dispersion by reducing particle–particle attraction. Conversely, at higher concentrations, silicate anions promoted the early formation of calcium silicate hydrate (C-S-H), which bridged particles together, overpowering the dispersing effect and increasing the static yield stress.

While the in-line addition of alkaline activators to AAM slurry is promising, it also presents significant challenges, such as requiring a very high water-to-precursor ratio. For instance, the AAM slurry in Ref. [15] required a water-to-precursor ratio of 0.45 to achieve a pumpable consistency. However, after inline activation, the total water-to-precursor ratio increased to approximately 0.70, which adversely affects early-age hardened properties and long-term durability. Subsequent study addressed this issue by optimizing the silicate modulus and incorporating specific admixtures [16]. However, further research is required to ensure its adoption in large scale construction.

3 Thermal Induced Setting

3.1 Microwave Heating

Microwaves, high-frequency electromagnetic waves ranging from 300 MHz to 300 GHz, are an effective method for heating dielectric materials like concrete. When exposed to microwaves, dielectric materials heat up due to internal friction as molecular attraction forces hinder induced vibrations. This internal heating accelerate the hydration reaction and drastically reduces the setting time [18]. A well-designed microwave heating regime can increase the setting rate of concrete, while an improper regime may damage the internal structure, weakening the concrete [19]. Based on this knowledge, microwave heating has been investigated as a method for active setting control in 3D printable concrete [20]. In this approach, a mix with high open time is pumped to the print head (Fig. 3), where it is exposed to microwave radiation to accelerate setting. It was observed that the setting rate increased significantly with just a few seconds of microwave heating, making it an ideal heating source. Accelerated setting not only enhanced the buildability of the printed layers but also influenced the interlayer adhesion. It was found that interlayer bond strength improved with increasing microwave heating duration, from 0 to 10 s. However, further increases in heating duration, while accelerating the setting rate, resulted in reduced interlayer bond strength. This could be attributed to the excessive hardening of the printed layer, reducing its malleability and interlocking capability. Therefore, microwave heating at the print head can be utilized to achieve tailored setting rates for specific printing requirements. For example, shorter microwave heating durations could be utilized for structures with longer cycle times (the time interval between successive layers). Conversely, for structures with short cycle times or those requiring rapid setting, longer microwave heating durations can be applied.

Microwave heating has also been explored in precast construction and pavement repair applications, where rapid setting of concrete is critical. This technique shows significant promise for various construction scenarios requiring active control over setting rates [21]. Even though microwave heating significantly enhances productivity in construction applications, its widespread adoption may face resistance due to safety concerns. Microwave technology has been in use for a considerable time, particularly in domestic and industrial applications, such as the food industry. However, its implementation in the construction industry still requires further development, both in terms of equipment design and achieving social acceptance. This presents an intriguing domain for interdisciplinary research, where engineering advancements can be coupled with studies addressing public perception. Bridging these gaps could accelerate the integration of microwave heating into mainstream construction practices.

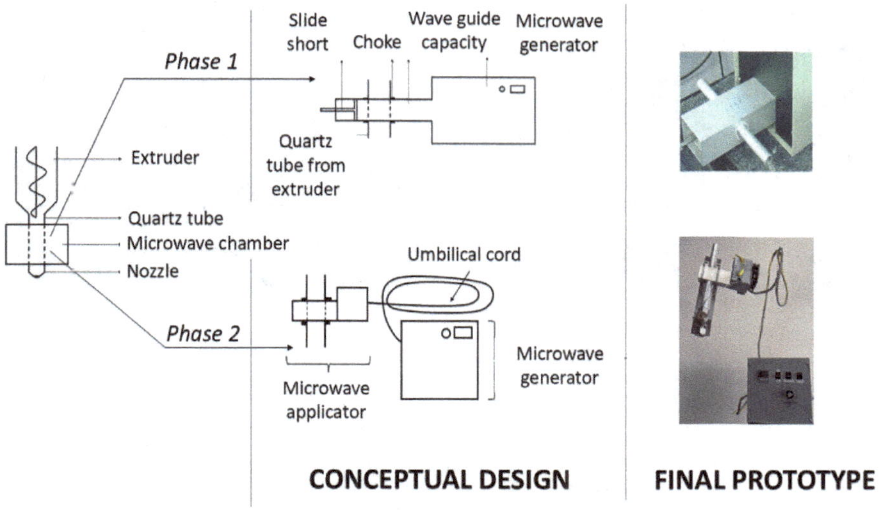

Fig. 3 Integrated print head with microwave to attain active control on the setting of the printed layers. Images were reproduced from Ref. [20]

3.2 Electric Heating

The application of electric current has emerged as an effective method to influence hydration kinetics and improve the mechanical properties of fresh mortar [22]. This technique (Fig. 4) shows potential for accelerating the setting rate. The application of electric current accelerates the hydration reaction by resistive (Joule) heating generated by the current [23]. Additionally, the application of current also increases the ion mobility, which enhances the dissolution of clinker phases and facilitates faster formation of hydration products further accelerating reaction kinetics [24]. According to Ref. [25], electric current application enhanced the formation of calcium silicate hydrate (C-S-H) gel in OPC paste, which contributed to the rapid setting of concrete.

Electric heating requires precise control, as higher-than-required current intensities can cause non-uniform hydration, potentially compromising long-term durability [26]. This could be mitigated with the application of alternating current (AC) instead of direct current (DC), as AC fields prevent the localized deposition of hydration products, maintaining uniform hydration [27]. Nevertheless, DC currents typically in the range of 1–10 V, have been reported to reduce the activation energy for hydration, accelerating setting times without adversely affecting long-term mechanical properties [27, 28]. Studies have shown that varying current intensities can precisely

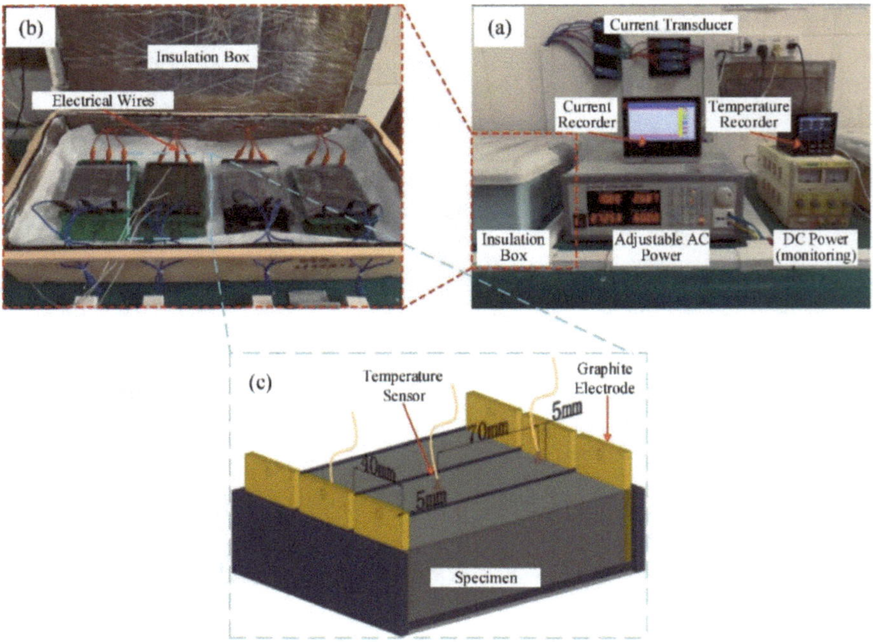

Fig. 4 Set up for electric curing system: **a** Data recorder **b** Samples in the insulation box **c** Thermal resistance temperature sensor in sample [23]

adjust the hydration rate [24, 29], enabling precise control over the setting behaviour during the construction process. Electric heating as an active setting control method can be promising for 3DCP. Compared to microwave heating (discussed earlier), the design and installation of a print head equipped for electric current application would be simpler. Additionally, this method is likely to encounter fewer challenges in obtaining work safety clearances compared to microwave heating [1]. However, ensuring consistent current application under variable field conditions is still crucial. Furthermore, a detailed understanding of the effects of electric current parameters on different cementitious compositions and printing scales is necessary for its effective implementation.

3.3 *Ultrasonic-Induced Setting*

In the field of dentinal cement, ultrasonic activation is widely used. This technique involves transmitting acoustic energy from an ultrasonic insert to a liquid, generating ultrasonic waves [30]. The process induces swirling of the irrigating solution, which creates fluctuations in hydrostatic pressure, resulting in the formation and implosion of cavitation bubbles. These implosions increase the material's temperature and effectively remove the smear layer. Consequently, ultrasonic activation promotes deeper penetration of the cement into the dentinal tubules, thereby enhancing the bond strength between the material and the root dentin. De Carvalho et al. [31] investigated the setting times of three different dental repair cements subjected to ultrasonic activation during mixing at a frequency of 24–32 Hz. The study revealed that ultrasonic activation reduced the setting time by approximately 2 min for the tested samples. This reduction aligns with enhanced reaction kinetics facilitated by ultrasonic waves. Moreover, ultrasonic waves promote particle dispersion, increasing the contact surface area for hydration reactions. Conversely, Lopes et al. [32] reported no significant influence of ultrasonic activation (20 s) on the setting time of calcium silicate sealers, highlighting potential variability in the effect of ultrasonic activation based on material composition.

Power ultrasound (PUS) is a technique that generates sound waves in the frequency range of 20–100 kHz and delivers energy levels exceeding 10 W/cm^2. PUS has been widely utilized across various industrial applications, particularly for processes such as deagglomeration and cleaning [33]. Rößler et al. [34, 35] developed a flow-through cell system that maintained a constant frequency of 20 kHz, with adjustable amplitudes, to investigate its effect on hydration reactions. Their findings indicated that PUS accelerated cement hydration and setting rate, a result attributed to enhanced alite hydration. Notably, faster setting times were also observed in mortars, even those with low clinker content [36, 37]. Additional in-depth studies on the impact of PUS on the hydration kinetics of C3S and the precipitation of C-S-H can be found in [38].

In the context of 3D concrete printing, Vaitkevičius et al. [39] utilized an ultrasonic converter integrated into the print head to control the hydration reaction during the 3D printing process. Their study demonstrated that ultrasonic activation, with dispersion times ranging from 0 to 30 s, significantly accelerated the hydration process, reducing the setting time by approximately four times. The authors explained that the ultrasonic activation generates liquid jets that travel at speeds of up to 1000 km/h. These high-speed jets exert substantial pressure on the liquid between the particles, causing them to separate and increasing the surface area available for reaction. Additionally, the acceleration of particles by the liquid jets leads to high-speed collisions, which generate heat and further accelerate the hydration process. The study revealed that longer dispersion times enhanced the hydration reaction, thereby reducing the setting rate. This allows for active control over the setting time of the printed layers by adjusting the duration of ultrasonic activation.

Similar results were reported by Serelis et al. [40], who examined the impact of varying ultrasonic activation times on the setting behaviour of cement pastes. They observed that extending the ultrasonic dispersion from 0 to 120 s resulted in a reduction of both the initial setting time, from 371 to 291 min, and the final setting time, from 754 to 615 min. On the contrary, Ehsani et al. [41] treated cement pastes with a water-to-cement (w/c) ratio of 0.30 using ultrasound at frequencies of 26, 67, and 132 kHz for a duration of two minutes and found delayed initial and final setting times. The authors attributed the outcomes to the disturbances in the hydration of aluminate and alite during ultrasound exposure.

4 Encapsulation of Chemical Additives

The controlled release of capsules using internal or external signals, has been widely applied in drug delivery and food industry [42, 43]. Also, for concrete different types of capsules incorporating phase change materials, inhibitors, and healing ingredients have been used in order to improve physical and durability properties [44, 45]. As a consequence of the increasing importance of the setting behaviour of cement-based materials, chemical admixtures were subjected to encapsulation to adjust the cement hydration. Choi et al. [46] presented a technique to encapsulate both an accelerator and a retarder. The study demonstrated that encapsulating lithium carbonate (accelerator) and tartaric acid (retarder) enabled the time-dependent release of these additives. This innovative approach effectively controlled the setting rate of concrete by regulating the sequential availability of the accelerator and retarder during the hydration process. It was further observed that the release rate of the additives was primarily governed by the wall thickness of the capsules, which were made from a soluble acrylic resin-based compound.

More recently, Shao et al. [47] developed a method to actively open microcapsules, measuring 2–4 mm in diameter, filled with chemical admixtures such as hydroxypropyl methylcellulose (HPMC) to regulate the setting behaviour of cement pastes (Fig. 5). The wall material of the capsules, composed of polyethylene glycol (PEG2000), was embedded with graphite powder, which acted as a microwave-responsive component. Microwaves at a frequency of 2450 MHz $\pm$ 50 Hz and a power of 600 W were employed to stimulate capsule opening. Upon the release of the admixtures, a significant increase in static and dynamic yield stress, as well as plastic viscosity, was observed. Similar method was used by Kanagasuntharam et al. [48], where thermal activation of capsules was achieved by circulating hot water around the print head to heat the capsules. The capsules utilized a thermo-responsive paraffinic phase change material with a melting temperature of 31 °C as the wall material, while sodium silicate served as the activator. Their findings indicated that the encapsulated sodium silicate significantly enhanced buildability compared to other thixotropic additives such as nano clay. In a subsequent study, the same authors encapsulated sodium hydroxide (NaOH) using a phase change material [49]. Cement mixtures containing these microcapsules were heated at the print head, resulting in rapid concrete setting upon placement. However, the NaOH additive adversely affected the long-term properties of the concrete. In the context of alkali-activated materials, Sun et al. [50] developed a method to encapsulate activator solutions to influence the setting time. Their research demonstrated that encapsulating sodium silicate extended the setting times by 40%. For the shell material, a cellulose-based gelatinous compound was used. However, the release mechanism was not actively controlled; the encapsulated sodium silicate was introduced during the initial mixing stage, with the capsule shells dissolving completely within 60 min of water addition, without the need for external stimuli such as microwave radiation or conventional heating.

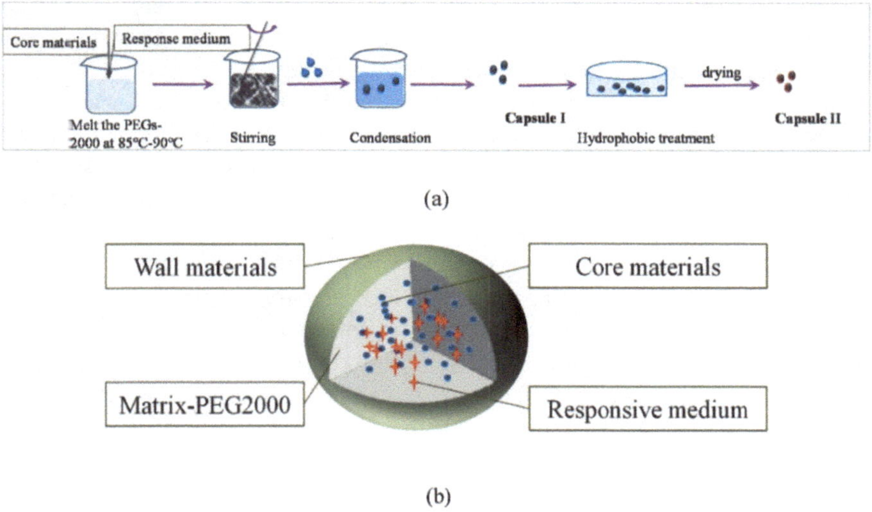

Fig. 5 Stepwise procedure on the preparation of capsules for active setting control. The figure was reproduced from Ref. [47]

5 Summary and Prospects

This chapter has explored various methods reported for achieving active control over the setting of concrete. Based on the reviewed literature, chemically induced setting has been widely utilized, particularly in modern construction techniques such as 3D concrete printing (3DCP). This approach typically involves the integration of an in-line mixer (static or dynamic) at the print head, which blends chemical admixtures with the base mix immediately before printing. While the focus has largely been on the mixture formulation (accelerate type, dosage etc.), the design of the in-line mixer has received comparatively less attention. Although faster setting rates can be achieved even with low mixing efficiency, inadequate mixing can compromise the long-term properties of the concrete. Consequently, optimizing the design of in-line mixers to maximize mixing efficiency for various concrete types represents a promising area for future research. When it comes to using heat as a signal for actively controlling the setting of concrete, microwave radiation and electric heating can offer significant advantages over conventional methods, particularly in achieving uniform heating with shorter heating durations. However, their adoption in the construction industry faces challenges related to work safety clearances, which present not only technical but also psychological barriers. Although microwave and electric heating technologies have been used safely in domestic applications for decades, there remains hesitation in applying them to construction processes. Therefore, interdisciplinary research focused on developing integrated print heads equipped with microwave or electric heaters, while simultaneously addressing social acceptance, represents a compelling area for future exploration.

The combination of two or more signals to achieve active control over the setting of concrete can be highly effective, as one signal can complement the other by addressing the respective challenges. For example, in the encapsulation of accelerators approach, setting can be actively controlled by utilizing both chemical and heat as stimulating signals. The primary challenge associated with chemically induced setting is the dependency on the mixing efficiency of the in-line mixer. This issue can be mitigated by encapsulating the chemical accelerator and ensuring its uniform dispersion during the initial mixing stage. The uniformly distributed encapsulated accelerators can then be released by activating the capsules with a secondary signal, such as heating, immediately before printing. Such complementary combinations of stimulating signals hold significant potential for enhancing control over concrete setting and represent an intriguing area for further investigation.

References

1. Muthukrishnan, S., Ramakrishnan, S., Sanjayan, J.: Technologies for improving buildability in 3D concrete printing. Cem. Concr. Compos. **122**, 104144 (2021)
2. Nicoleau, L.: Accelerated growth of calcium silicate hydrates: experiments and simulations. Cem. Concr. Res. **41**(12), 1339–1348 (2011)

3. Abdelrazig, B., et al.: The solution chemistry and early hydration of ordinary portland cement pastes with and without admixtures. Thermochim. Acta **340**, 417–430 (1999)
4. Wang, Y., et al.: Accelerators for shotcrete–chemical composition and their effects on hydration, microstructure and properties of cement-based materials. Constr. Build. Mater. **281**, 122557 (2021)
5. Hu, Z., et al.: Powder accelerated high volume slag cement for lower carbon sprayed concrete linings. Constr. Build. Mater. **449**, 138308 (2024)
6. Wangler, T., et al.: A chemical process engineering look at digital concrete processes: critical step design, inline mixing, and scaleup. Cem. Concr. Res. **155**, 106782 (2022)
7. Tao, Y., et al.: Blending performance of helical static mixer used for twin-pipe 3D concrete printing. Cem. Concr. Compos. **134**, 104741 (2022)
8. Wangler, T., et al.: Aluminate 2K systems in digital concrete: Process, design, chemistry, and outlook. Cem. Concr. Res. **185**, 107644 (2024)
9. Galli Rodrigues De Melo, F.: Service life of shotcrete: investigation on the effect of set accelerating admixture (2018)
10. Taylor, H.: Cement Chemistry. Thomas Telford (1997)
11. Jiang, Z., et al.: Evaluating the stiffening effect of CSA and sodium carbonate on the printability of OPC mortar. Constr. Build. Mater. **328**, 127088 (2022)
12. Tao, Y., et al.: Development of a calcium sulfoaluminate-Portland cement binary system for twin-pipe 3D concrete printing. Cem. Concr. Compos. **138**, 104960 (2023)
13. Tan, H., et al.: Effect of borax on rheology of calcium sulphoaluminate cement paste in the presence of polycarboxylate superplasticizer. Constr. Build. Mater. **139**, 277–285 (2017)
14. Marchon, D., et al.: Hydration and rheology control of concrete for digital fabrication: potential admixtures and cement chemistry. Cem. Concr. Res. **112**, 96–110 (2018)
15. Muthukrishnan, S., Ramakrishnan, S., Sanjayan, J.: In-line activation of geopolymer slurry for concrete 3D printing. Cem. Concr. Res. **162**, 107008 (2022)
16. Muthukrishnan, S., Ramakrishnan, S., Sanjayan, J.: Rapid early age strength development of in-line activated geopolymer for concrete 3D printing. Constr. Build. Mater. **406**, 133312 (2023)
17. Muthukrishnan, S., Ramakrishnan, S., Sanjayan, J.: Set on demand geopolymer using print head mixing for 3D concrete printing. Cem. Concr. Compos. **128**, 10445 (2022)
18. Xuequan, W., Jianbgo, D., Mingshu, T.: Microwave curing technique in concrete manufacture. Cem. Concr. Res. **17**(2), 205–210 (1987)
19. Li, T., et al.: Experimental study of microwave heating on mechanical properties of fly ash-based cementitious materials. J. Build. Eng. 108454 (2024)
20. Muthukrishnan, S., Ramakrishnan, S., Sanjayan, J.: Effect of microwave heating on interlayer bonding and buildability of geopolymer 3D concrete printing. Constr. Build. Mater. **265**, 120786 (2020)
21. Leung, C.K., Pheeraphan, T.: Microwave curing of Portland cement concrete: experimental results and feasibility for practical applications. Constr. Build. Mater. **9**(2), 67–73 (1995)
22. Susanto, A., Koleva, D.A., van Breugel, K.: The effect of water-to-cement ratio and curing on material properties of mortar specimens in stray current conditions. J. Adv. Concr. Technol. **15**(10), 627–643 (2017)
23. Yang, Z., et al.: Experimental investigation on mechanical strength and microstructure of cement paste by electric curing with different voltage and frequency. Constr. Build. Mater. **299**, 123615 (2021)
24. Uygunoğlu, T., Hocaoğlu, İ., Topçu, İ.B.: Electrical curing application on cement-based mortar with different stress intensity. In: Proceedings of 3rd International Sustainable Buildings Symposium (ISBS 2017), vol. 2. Springer (2018)
25. Ylmén, R., et al.: Early hydration and setting of Portland cement monitored by IR, SEM and Vicat techniques. Cem. Concr. Res. **39**(5), 433–439 (2009)
26. Lee, J.-C., Yi, C.-Y.: Setting process monitoring of cement paste using electromechanical impedance of piezoelectric patch. Materials **15**(22), 8114 (2022)
27. Tang, S., et al.: The review of early hydration of cement-based materials by electrical methods. Constr. Build. Mater. **146**, 15–29 (2017)

28. Hocaoğlu, İ: Self-cooling mortar production with zinc oxide nanoparticles additive and investigation of the DC application when early-age hydration. Eur. J. Environ. Civ. Eng. **27**(12), 3586–3602 (2023)
29. Uygunoğlu, T., Hocaoğlu, İ: Effect of electrical curing application on setting time of concrete with different stress intensity. Constr. Build. Mater. **162**, 298–305 (2018)
30. Van der Sluis, L., et al.: Passive ultrasonic irrigation of the root canal: a review of the literature. Int. Endod. J. **40**(6), 415–426 (2007)
31. Acris De Carvalho, F.M., et al.: Influence of ultrasonic activation on the physicochemical properties of calcium silicate-based cements. Int. J. Dent. **2021**(1), 6697988 (2021)
32. Lopes, F.C., et al.: Effect of sonic and ultrasonic activation on physicochemical properties of root canal sealers. J. Appl. Oral Sci. **27**, e20180556 (2019)
33. Suslick, K.S., Price, G.J.: Applications of ultrasound to materials chemistry. Annu. Rev. Mater. Sci. **29**(1), 295–326 (1999)
34. Rößler, C., et al.: Power ultrasound—an effective method to accelerate setting and early strength development of concrete. In: 13th International Congress on the Chemistry of Cement, Madrid, Spain (2011)
35. Peters, S.: The influence of power ultrasound on setting and strength development of cement suspensions (2017)
36. Remus, R., et al.: Power ultrasound assisted production of sustainable concrete. In: 16th International Congress on the Chemistry of Cement 2023 (ICCC2023), Bangkok, Thailand (2023)
37. Remus, R., Roessler, C., Ludwig, H.M.: Power ultrasound-assisted concrete production—workability, strength development, and durability, pp. 135–149. ACI Special Publication, American Concrete Institute (2018)
38. Remus, R., et al.: Fundamental effects of using power ultrasound to accelerate C3S hydration. Cem. Concr. Res. **180**, 107514 (2024)
39. Vaitkevičius, V., Šerelis, E., Kerševičius, V.: Effect of ultra-sonic activation on early hydration process in 3D concrete printing technology. Constr. Build. Mater. **169**, 354–363 (2018)
40. Serelis, E., et al.: Effect of ultra-sonic dispersion time on hydration process and microstructure development of ultra-high performance glass powder concrete. Constr. Build. Mater. **298**, 123856 (2021)
41. Ehsani, A., et al.: Insights into the positive effects of power ultrasound on the pore solution of Portland cement pastes. Cem. Concr. Compos. **125**, 104302 (2022)
42. Rivas, C.J.M., et al.: Nanoprecipitation process: from encapsulation to drug delivery. Int. J. Pharm. **532**(1), 66–81 (2017)
43. Gibbs, F., Kermasha, S., Alli, I., Catherine N. Mulligan, B.: Encapsulation in the food industry: a review. Int. J. Food Sci. Nutr. **50**(3), 213–224 (1999)
44. Souradeep, G., Kua, H.W.: Encapsulation technology and techniques in self-healing concrete. J. Mater. Civ. Eng. **28**(12), 04016165 (2016)
45. Hunger, M., et al.: The behaviour of self-compacting concrete containing micro-encapsulated phase change materials. Cem. Concr. Compos. **31**(10), 731–743 (2009)
46. Choi, Y.C., et al.: Development and application of microcapsule for cement hydration control. KSCE J. Civ. Eng. **20**, 282–292 (2016)
47. Shao, L., et al.: A novel method for improving the printability of cement-based materials: controlling the releasing of capsules containing chemical admixtures. Cem. Concr. Compos. **128**, 104456 (2022)
48. Kanagasuntharam, S., Ramakrishnan, S., Sanjayan, J.: Encapsulation of sodium silicate to attain on demand buildability enhancement in concrete 3D printing. Jo. Build. Eng. 109912 (2024)
49. Kanagasuntharam, S., Ramakrishnan, S., Sanjayan, J.: Investigating PCM encapsulated NaOH additive for set-on-demand in 3D concrete printing. Cem. Concr. Compos. **145**, 105313 (2024)
50. Sun, M., et al.: Innovative encapsulation of alkali activators in alkali-activated slag concrete: a sustainable strategy for regulating setting time and durability. Constr. Build. Mater. **427**, 136230 (2024)

Active Control of Concrete Shrinkage and Associated Cracking

William Peter Boshoff, Eleni Korda, Didier Snoeck, and Gashaw Abebaw

Abstract Concrete shrinkage is a significant concern in construction due to its potential to cause non-structural cracking. The causes of shrinkage during the setting and hardening of concrete are multifaceted, involving both physical and chemical mechanisms. These phenomena are influenced by various factors, including environmental conditions, material properties, curing techniques, and mix design. The control of shrinkage properties in concrete is vital for minimising the risk of cracking. Advances in material innovation and processing techniques have demonstrated that adjusting the cement matrix, incorporating expansive additives, and using shrinkage-reducing admixtures can significantly reduce shrinkage. Furthermore, internal curing methods, such as the use of pre-saturated lightweight aggregates and super-absorbent polymers, have proven effective in mitigating autogenous shrinkage. In this chapter, key studies on autogenous and plastic/drying shrinkage are reviewed, highlighting their mechanisms, and possible methods for active control to optimise concrete properties. The active control methods are further explained in chapter "Active Control of Cementitious Slabs and Roads".

Keywords Active control of concrete · Shrinkage · Cracking · Concrete · Autogenous shrinkage · Plastic shrinkage

W. P. Boshoff (✉)
University of Pretoria, Pretoria, South Africa
e-mail: billy.boshoff@up.ac.za

E. Korda
Vrije Universiteit Brussel, Brussels, Belgium

Ghent University, Ghent, Belgium

D. Snoeck
Université Libre de Bruxelles, Brussels, Belgium

G. Abebaw
Debre Tabor University, Debre Tabor, Ethiopia

G. De Schutter and J. Sanjayan (eds.), *Active Control of Properties of Fresh and Hardening Cementitious Materials*, RILEM State-of-the-Art Reports 50,
https://doi.org/10.1007/978-3-032-30340-0_5

1 Introduction

Concrete shrinkage is a significant concern in construction due to its potential to cause non-structural cracking and related durability issues. The causes of shrinkage during the setting (fresh state) and hardening of concrete are multifaceted, involving both physical and chemical mechanisms. These phenomena are influenced by various factors, including environmental conditions, material properties, curing techniques, and mix design amongst others.

Controlling shrinkage properties in concrete is vital for minimising the risk of cracking. This can lead to a severe reduction in service life or increase the need for maintenance if not controlled. Advances in material innovation, curing methods and processing techniques have demonstrated that adjusting the cement matrix, incorporating expansive additives, and using shrinkage-reducing admixtures can significantly reduce shrinkage. Furthermore, internal curing methods, such as the use of pre-saturated lightweight aggregates and super-absorbent polymers, have proven effective in mitigating autogenous shrinkage and associated cracking.

In this chapter, key studies on autogenous and plastic/drying shrinkage are reviewed, highlighting their mechanisms, and possible methods for active control to optimise concrete properties. The active control methods are further explained in chapter " Active Control of Cementitious Slabs and Roads".

2 Chemical Shrinkage

Chemical shrinkage is the internal volume change caused by the chemical reaction between cement and water [1] when the concrete is still in a fresh state and before setting. This normally happens within the first few hours after the cement has come in contact with water. This can strictly happen whether the concrete surfaces are exposed to the environment or not. Chemical shrinkage is, however, most often measured in sealed conditions. The measured shrinkage in sealed conditions is known as autogenous shrinkage even though the sealed conditions are not a prerequisite for chemical shrinkage.

Chemical shrinkage is a major factor causing cracking in early-age concrete. Chemical shrinkage occurs when the volume of hydration products is smaller than the combined initial volume of the reacting components, namely cement and water. It is defined as the reduction in internal volume within the hydrated cement paste of concrete [2]. In the early stage, when concrete remains viscous, autogenous shrinkage is primarily caused by chemical changes resulting from cement hydration. However, after the initial set, it is mainly driven by self-desiccation, which is impaired by specific mix components, such as silica fume, superplasticizers, and a low water/cement ratio [2]. Consequently, high-strength and high-performance concrete are at a greater risk of autogenous shrinkage. Autogenous shrinkage is affected by factors such as the water/cement ratio, cement composition, temperature, relative humidity, and/or the inclusion of supplementary cementitious materials. A higher water-to-cement ratio generally decreases autogenous shrinkage as more water is available for hydration, especially when comparing to a water-to-cement ratio of 0.3 [3]. Additionally, elevated temperatures and reduced relative humidity can speed up the shrinkage process.

Chemical shrinkage is proportional to the degree of hydration. Compositions with higher water/binder ratios, will achieve a somewhat higher ultimate degree of hydration, and, therefore, undergo a somewhat higher chemical shrinkage compared to lower water/binder ratios [4]. Finally, chemical shrinkage in pastes increases with a higher water/binder ratio and temperature [5]. Nevertheless, because of increased self-desiccation, the autogenous shrinkage is higher in case of lower water/binder ratios. The volume of paste also plays a role if aggregates are added.

As aforementioned, cement and water react to form crystalline and gel hydration products. As a result, chemical shrinkage occurs as the products occupy less space than the reactants, leading to a non-negligible volume reduction. A generally accepted reaction stoichiometry for the hydration of tricalcium silicate being the major component of Portland cement is:

$$\begin{array}{cccc} C_3S + 5.3H & \rightarrow & C_{1.7}SH_4 + 1.3CH \\ 71.1 \quad 95.8 & & 107.8 \quad 43 \end{array} \tag{1}$$

When considering the molar volumes shown below the equation, the hydration of the tricalcium silicate results to a net reduction of 9.6% of volume or approximately 0.07 mL/g C_3S, meaning 7% of water mass needed per mass of C_3S [6].

Le Chatelier measured the volume reduction in cement paste by means of a simple dilatometer using a glass beaker attached to a tube filled with water. It was found that the cumulative chemical shrinkage of Portland cement was 4.60% by weight of cement [7]. To determine chemical shrinkage, the cement paste must remain water-saturated, ensuring that external imbibed water replaces the empty pores responsible for the total volume change. Finally, chemical shrinkage is usually up to two times higher than autogenous shrinkage and is directly dependent on the degree of hydration [4].

Chemical shrinkage development is strongly regulated by cement clinker concentration and type. The higher the C_3A and C_4AF content, the larger the chemical shrinkage, and the alkaline content increases it [8]. Cement grain dissolution volume changes and water consumption through hydration reactions are the mechanisms causing chemical shrinkage [9]. Cement fineness may be the most significant factor influencing chemical and autogenous shrinkage in cement paste, as it determines the particle surface area in contact with water [1, 10]. As seen in Fig. 1, autogenous shrinkage occurs when cement paste undergoes a hydration process that results in a volume change and the emergence of new products.

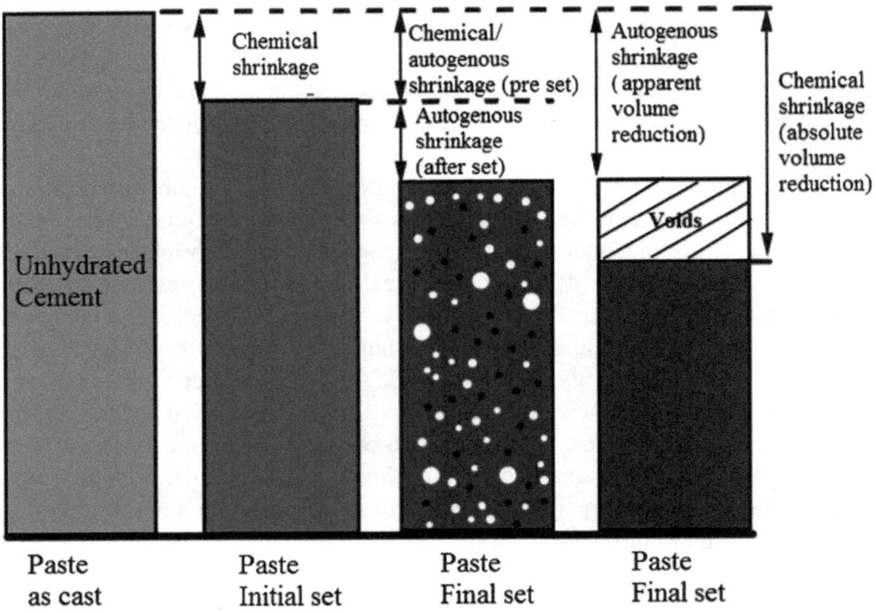

Fig. 1 Autogenous and chemical shrinkage of cement paste [11]

3 Self-desiccation

Self-desiccation is often wrongly used interchangeably with chemical shrinkage and autogenous shrinkage. Self-desiccation is however a significantly different mechanism to chemical shrinkage and is typically measured in sealed conditions, thus making out part of the measured autogenous shrinkage as defined in the previous sub-section. Autogenous shrinkage can also be called apparent volume reduction when concrete is sealed.

The mechanism of self-desiccation is the drying of the relatively fresh concrete from the inside due to the cement hydration reaction. This drying causes a significant shrinkage that cannot be ignored. This drying can be aggravated or mitigated by concrete not being sealed for moisture interaction with the environment. The aggravation comes from the concrete losing even more water through drying to the environment while self-desiccation can be mitigated by keeping the concrete in a high relative humidity/wet environment during curing. The higher the strength of the concrete, i.e. in general, the lower the water/binder ratio, the less the mitigating effect. This is due to the permeability of higher strength concrete being significantly lower, thus water struggles to penetrate the dry concrete to replenish the water used by the cement reaction. This is one of the reasons why lower water/binder ratio concrete exhibit more self-desiccation, thus higher measured autogenous shrinkage.

The self-desiccation of High Performance Concrete (HPC) helps mitigate moisture-related issues during construction. In conventional concrete, excessive moisture can lead to detachment of glued plastic carpets, buckling of wood floors, but also to durability issues. With self-desiccation, the internal relative humidity rapidly drops below the critical level, reducing this risk [12]. However, the low permeability of HPC restricts water ingress into the interior of large structures. As a result, self-desiccation can lead to autogenous shrinkage, potentially causing cracking, making, therefore, the control of self-desiccation crucial for mitigating the latter. Self-desiccation causes HPC to shrink even when submerged in water, unlike conventional concrete, which shrinks only during drying when exposed to air. This is due to the fact that in HPC, internal drying occurs due to the chemical shrinkage of water as it binds to cement during hydration [12].

Both chemical shrinkage and self-desiccation lead to volume reductions. Before setting, chemical shrinkage is responsible for the volume reduction of the microstructure. However, after final setting, when the cementitious matrix has developed its resistance to deformation, the chemical shrinkage in absence of additional water source will lead to self-desiccation due to partially water-filled pores being produced [13]. The menisci formed in the pore structure (capillary pores) are responsible for the development of the so-called capillary pressure force, which is directly proportional to the surface tension of the pore solution and inversely proportional to the largest capillary pore size, as explained in the capillary pressure section below. This capillary pressure will result in a measurable shrinkage of the cementitious matrix.

Another driving factor for self-desiccation is the hydration reaction itself. As hydration proceeds, self-desiccation is enhanced, causing the larger pores to empty first, as the water inside the paste is consumed by chemical reactions (hydration of cement particles). This process leads to a reduction in internal relative humidity, particularly in the smaller pores, making the matrix porosity another important factor. As a result, the capillary forces generated by the decrease in moisture content can lead to additional shrinkage. Self-desiccation typically occurs in the early stages of hydration, especially in low water/binder ratio mixes, and is especially prevalent in high-performance concrete mixes. While it is also driven by capillary action, self-desiccation is more focused on the loss of water due to hydration and the resulting increase in internal suction pressures. The main driving mechanisms for self-desiccation are reviewed in the following sections after which examples of typical testing methods are shown.

3.1 Capillary Forces

The ongoing hydration causes the amount of free water in the matrix to decrease over time, and, hence, the internal RH. As a result, numerous pores form in the hardened cement paste, and the water saturation within these pores declines. As capillary pores transition from a saturated to an unsaturated state, the concave inner surface of the pore experiences internal pressure. To bring the concave surface into a state of equilibrium, the capillary tension increases, resulting in self-desiccation and, therefore, shrinkage [14].

The difference between the gas pressure above the meniscus and pressure inside the liquid is called capillary pressure, and it can be expressed by the Young–Laplace equation [15] as follows:

$$P_c = P_g - P_l = -\frac{2\gamma \cos\theta}{r_c - t} \tag{2}$$

where P_c is the capillary pressure [Pa], P_g is the gas pressure [Pa], P_l is the liquid pressure [Pa], r_c is the capillary radius at the position of the meniscus [m], γ is the surface tension between the pore water and the vapor [N/m], θ is the contact angle that indicates the hydrophilicity of the pore wall [°], and t is the thickness of the adsorbed layer [m], which is a function of RH [16–18].

Capillary pressure can also be expressed by Kelvin's equation [19], which links the capillary pressure to the internal RH.

$$P_c = \rho_l RM^{-1}T\ln\frac{RH}{x_w} \tag{3}$$

where ρ_l is the density of the liquid [kg/m^3], M is the molar mass of the liquid [kg/mol], $R = 8.314$ [J/(mol K)] is the universal gas constant, T is the temperature [K], and x_w accounts for the presence of ions in the pore solution, based on Raoult's law, for example.

The Kelvin-Laplace equation shows that the Kelvin radius is mainly determined by RH and pore solution properties rather than the pore size. Since pores of various sizes are interconnected, any difference in meniscus radius before equilibrium leads to a capillary pressure gradient, driving water movement until equilibrium is reached. As the RH decreases, the meniscus radius shrinks in all pores, aiming for a new Kelvin radius. Once the Kelvin radius becomes smaller than the pore radius (excluding adsorbed layers), water in those pores drains as the meniscus cannot shrink further [20]. This capillary pressure gradient drives liquid from larger to smaller pores, with larger pores emptying first, as shown in Fig. 2. When equilibrated at a specific RH, smaller surface pores remain filled with water, while larger pores dry out, retaining only adsorbed water [21, 22].

3.2 *Pore Structure*

The terminology pore structure is typically used to describe porosity, pore size distribution and pore morphology. All aforementioned have a great influence on self-desiccation shrinkage. Different pore sizes will result in different generated shrinkage stresses. The migration of moisture from saturated to unsaturated pores is directly affected by the pore connectivity and is another major factor influencing shrinkage [23]. In HPC, a more refined pore network leads to a faster reduction in the hydraulic radius of capillary pores, intensifying self-desiccation [23]. Meanwhile, the

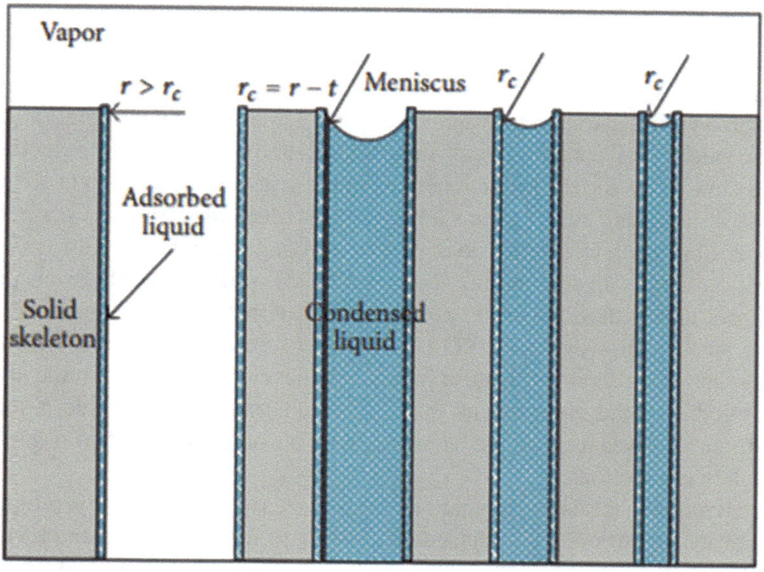

Fig. 2 Schematic illustration of drying process with different pore sizes, the formation of meniscus and generation of capillary pressure [20]

internal water consumption is increased by the chemical shrinkage, inducing further surface tension which increases capillary pressure [11]. Supplementary cementitious materials (SCM) such as fly-ash, silica fume and slag will affect the pore structure in different ways. When the water/binder ratio is kept constant (e.g. 0.3), fly-ash reduces autogenous shrinkage [24], while silica fume and slag increase it [25]. More specifically, silica fume and slag improved the micro-pore structure by reducing porosity and pore diameter, while increasing specific surface and the volumetric percentage of pores between 5 and 50 nm. This refinement increased capillary tension, leading to higher autogenous shrinkage [11].

3.3 Relative Humidity

Relative Humidity (RH) is defined as the ratio of the ambient vapor pressure to the saturation vapor pressure at the ambient temperature. In concrete, RH (or internal relative humidity) is a property of the air neighbouring the liquid water held in the capillary pore [26]. An interface of air and water exists within each pore, which forms a meniscus. Evaporation rate and local RH are a function of the restraining forces of the meniscus' curvature that hold the water molecules in the liquid water. When water is removed from the pores, either through external drying or the hydration reaction itself, the larger pores empty first, and the RH is reduced. In drying systems, environmental conditions play a crucial role in the evolution of the internal RH of concrete. RH drops with the increase in temperature when no moisture is added to the air. However, under sealed conditions, increasing the temperature will result in the expansion of the water, reduction of the meniscus curvature and the surface tension in the pore system, and therefore an increase in internal RH and evaporation rate.

Independently of concrete being in sealed or drying conditions, reductions in RH will slow down the hydration reaction, and practically terminate it at RH levels below 80%, since self-desiccation cannot further reduce the RH in the pores below that level. For concrete exposed to drying at early ages the drop in internal RH is even more crucial for determining their mechanical properties. RH levels can be maintained, and hydration can be prolonged if external curing is provided to the concrete surface. However, in HPC of low water-to-cement ratio and a fine pore structure, self-desiccation will cause internal RH levels to drop even in the absence of moisture loss to the environment. In this sense, internal curing provides a solution to the aforementioned mechanism by maintaining the pores saturated, as is discussed in the following sections.

The reduction in RH is a combination of various reasons. As hydration progresses, the free water in the matrix is consumed, leading to internal drying and therefore, increased capillary pore pressure. This makes the water/binder ratio a highly influential factor in the RH evolution [27, 28]. The pore structure is another factor influencing the internal RH, with the main mechanism driving RH changes being the number of saturated pores [29, 30]. RH levels are stable (at 98%) up to 12 h, until the final set of cement and when the system is still saturated, and drop subsequently, due to self-desiccation as shown in Fig. 3a [31]. The initial drop of 2% is, however, attributed to the dissolved salts in the pore fluid [32] and can be, therefore, better linked to chemical shrinkage. Once the two values of RH are separated, the Kelvin radius can be calculated if the RH is measured. The results are shown in Fig. 3b.

In sealed systems, from the thermodynamics perspective, progressive cement hydration reduces evaporative water in capillary pores, leading to a decrease in RH in concrete [33, 34]. Finally, a lower water/binder ratio and higher silica fume content leads to a decrease in the internal relative humidity of HPC. Additionally, the outer concrete layers exhibit higher RH and maturity values compared to the inner ones [35], a factor that can enhance autogenous shrinkage in HPC with low w/c ratios and concrete with supplementary cementitious materials (SCM). Another study demonstrated that the reduction of RH in HPC is not due to shortage of water, but because the evaporable water in HPC cannot evaporate freely [36]. More specifically, the authors attributed the reason behind self-desiccation to the increase in mole concentration of soluble ions and the reduction of pore size or the increase of micro-pore water in the total evaporable water.

3.4 Hydration Reaction

As hydration progresses, more water is bound, reducing the RH and therefore, inducing more self-desiccation. In HPC, due to the fineness of the microstructure,

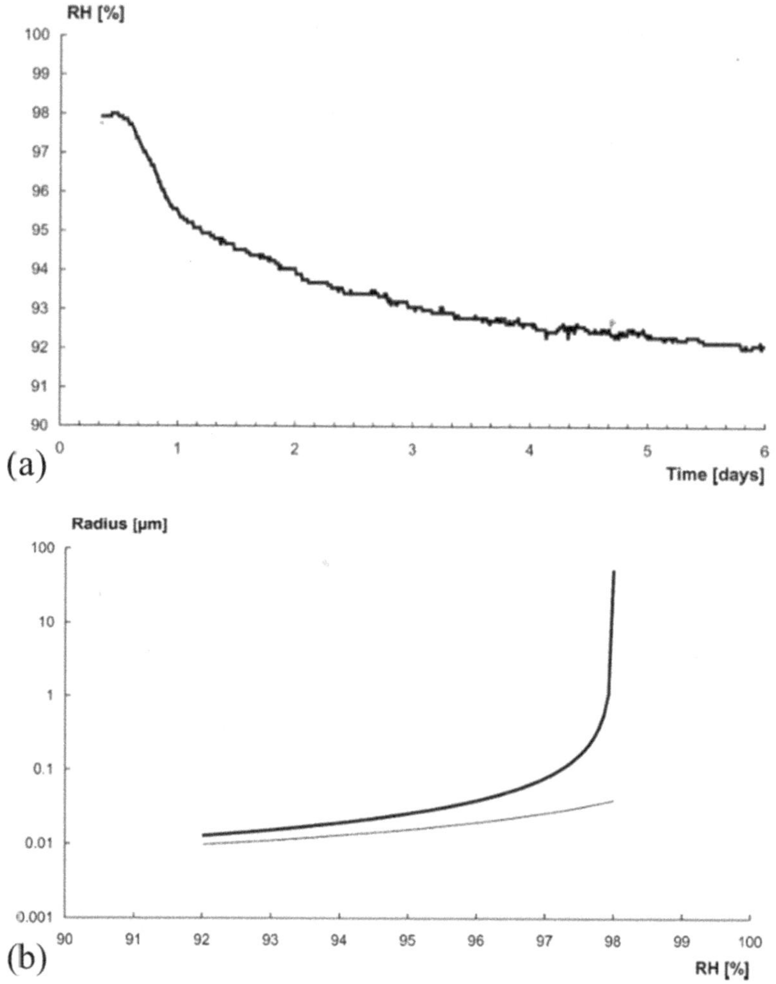

Fig. 3 a Autogenous internal RH measured on cement paste. Temperature 20 ± 0.1 °C. Average of two samples. **b** Kelvin's radius calculated with (bold line) and without (thin line) correction for the influence of dissolved salts. The RH drop due to salts (2%) is assumed to be constant [31]

self-desiccation due to the chemical reactions causes desaturation of the pores, formation of the menisci and the internal RH decreases even without loss of moisture to the environment [37]. Meanwhile, in theory, cement hydration slows significantly when RH drops below approximately 80% [38]. As a result, a lower degree of hydration is reached in low w/c ratio systems. The reason behind the latter is that self-desiccation alone cannot reduce RH levels in the pores below RH of 80% [39].

The reduced hydration in lower RH can be attributed to several mechanisms. Powers suggested that the desaturated pores and absence of water will not allow hydration products to be formed [38]. Other researchers have attributed the phenomenon to hydration kinetics, suggesting that the induction period of hydration is prolonged in reduced RH and, therefore, hydration onset is delayed [40]. From the thermodynamics point of view, the action of capillary tension is responsible for the reduced water activity terminating the alite hydration below 80% RH [41]. More specifically, the reduced water movement shifts the solubility of alite to keep it balanced with the pore fluid. The effect of RH on hydration kinetics in the so-called open cementitious systems (saturated with various thicknesses of water layers on top) and sealed systems (dry, no water layer on top) of low water/cement ratios was investigated in literature [37]. The sealed samples showed considerably lower rates of hydration when the RH was reduced due to self-desiccation. Apart from RH, the degree of hydration is highly influenced by the cement type and content, the presence of SCM and the temperature. Microstructural analysis or thermodynamics have been applied to model the degree of hydration degree as well as the C-S-H formation [42, 43].

3.5 Water/Cement Ratio

The water/cement ratio has a direct influence on the RH levels, and therefore, self-desiccation. Autogenous shrinkage is more severe in low water/cement ratio compositions (<0.4). From a water/cement ratio of 0.35–0.30, the jump in autogenous shrinkage seems to be significant, while a 10% addition of silica fume can enhance the same trend [44]. Other studies have shown that a reduction in water/binder ratio decreases RH of the cement slurry, especially at early ages, while significantly increasing self-desiccation, while further reduction in RH was attributed to the hydration reaction itself and was independent of the water/cement ratio [45].

Meddah [23] concluded that, for the same binder type and given age, increasing curing times and lower water/binder ratios resulted in lower total porosities and a finer pore structure. When using SF as partial replacement of Portland cement, they found that by lowering the water/binder ratio by only 5% (from 0.35 to 0.30), it becomes more effective in refining the pore structure than prolonging the curing time. However, the autogenous shrinkage increased at a higher rate with decreasing water/binder ratio, during the first two days of curing, while for water/binder ratio of 0.30, autogenous strains seemed to stabilise more quickly compared to concrete with water/binder of 0.35.

In unsealed systems, a higher water/cement ratio increases bleeding water, providing longer protection against evaporation and self-desiccation. The evaporation rate depends on temperature, pore solution properties, vapor pressure gradient, airflow rate, and diffusion coefficient. Evaporation from the drying surface drives liquid flow and vapor diffusion within the cementitious material [46, 47].

3.6 *Characteristics of Cement*

The characteristics of the utilised cement will greatly influence the self-desiccation process. Major constituent compounds include tricalcium silicate (C_3S), tricalcium aluminate (C_3A), dicalcium silicate (C_2S) and tetra-calcium aluminate ferrite (C_4AF). Some of the shrinkage-inducing mechanisms include the fineness of the cement (through the Blaine value), the amount of free lime in the clinker, which generates swelling [7], the amount of C_3A which increases autogenous deformation [48], and the sulphate-to-alkali ratio in the clinker which controls the C_3A crystalline form. Specifically, when SO_3/K_2O ratio is higher than 1, K_2O is mainly combined with SO_3. Tazawa [49] showed that autogenous shrinkage can be predicted by the cement composition. Initially, the rapid increase in heat of hydration and rapid formation of C_3S and C_3A creates a high temperature gradient in concrete, making it highly prone to shrinkage cracking at early age. The latter is followed by the subsequent consumption of the interlayer water and further volume reduction in the microstructure during the transformation process, which leads to greater shrinkage [50].

Regarding the impact of cement type on self-desiccation, the influence of C_3A seems to have the greatest influence, followed by C_3S, C_4AF and C_2S [51]. The higher the C_3A and C_3S contents, the higher the autogenous shrinkage in high-strength cementitious systems, due to the higher evolving temperature, as opposed to lower generating cements of higher C_2S contents [11]. A higher cement fineness and therefore increased specific surface area, results in faster water consumption, and RH decrease, and corresponding self-desiccation [52]. A higher cement fineness leads to a higher hydration rate up to 3 days of curing, causing a higher contraction phase of C-S-H gel. Autogenous shrinkage is high at this stage, slows down afterwards, and stops after 14 days of curing [53, 54]. A solution for HPC might be the use of coarse cement, which can absorb water for a longer period, leading to an increased long-term degree of hydration and reduced autogenous shrinkage according to [55]. Finally, the morphology of cement particles should be carefully considered. Spherical cement particles have a smaller contact area and a larger pore radius, which significantly helps reduce autogenous shrinkage in concrete [56].

3.7 *Disjoining Pressure*

In regions of restricted adsorption, where the gap between solid surfaces is less than twice the thickness of the freely adsorbed water layers, disjoining pressure becomes active [31]. This pressure arises from van der Waals forces, double-layer repulsion, and structural forces [57]. The effect is particularly significant at high relative humidity (RH) due to the abrupt change in the number of adsorbed water layers [58]. Disjoining pressure depends on RH levels and the concentration of $Ca2^+$ ions in the pore solution. Figure 4 shows that during the initial stage, the preexisting disjoining pressure in the pores is equilibrated by spreading pressure which is applied longitudinally to the solid skeleton interface. As RH decreases and water evaporates, the pre-exerting force decreases, releasing disjoining pressure, as well as reducing the distance between the two surfaces through shrinkage (Fig. 4) [20]. The loss of disjoining pressure is inherently linked to the desorption of water molecules and an increase in solid surface free energy. Additionally, if a capillary meniscus forms near the nanopores, it can further influence disjoining pressure [59].

Disjoining pressure is created in narrow locations, which have been defined as pores <0.6 nm wide [60]. When repulsive forces decrease, the distance between the solid particles also decreases, and shrinkage is formed. Moreover, the larger proportion of physically adsorbed water affects in turn the capillary tension in the pore structure [20]. Relationships have been proposed between changes in RH and

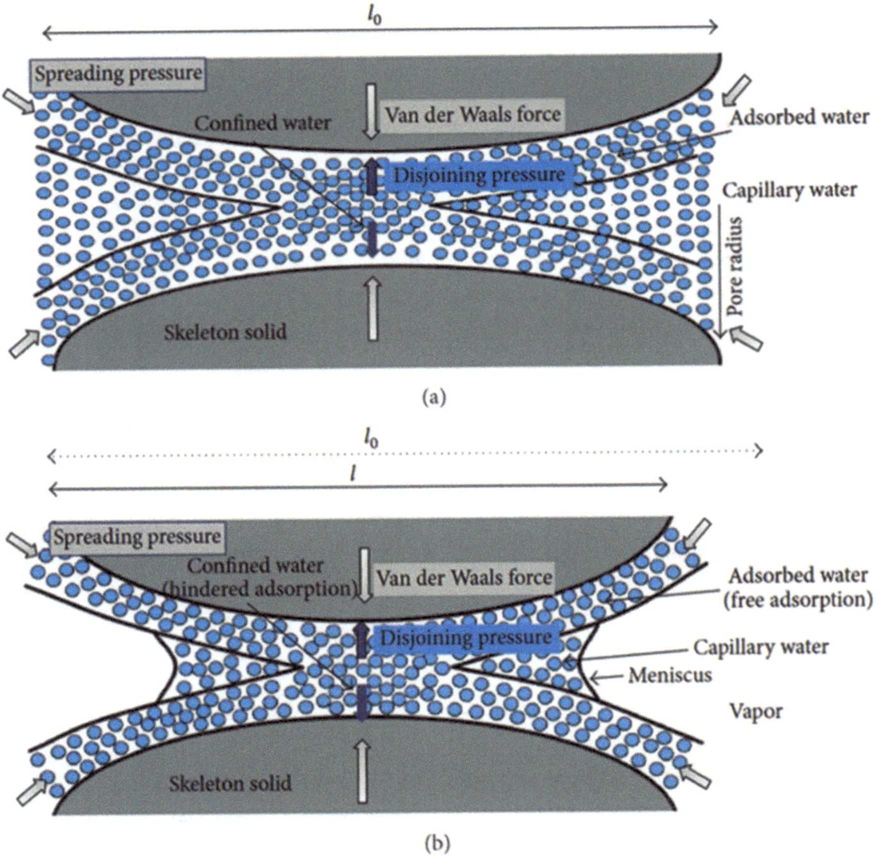

Fig. 4 Schematic illustration of drying process in hindered adsorption region and withdrawal of disjoining pressure: **a** initial status, **b** after RH decreases. This figure shows the simultaneous occurrence of surface free energy-induced shrinkage and withdrawal of disjoining pressure-induced shrinkage [20]

disjoining pressure on the deformation of hardened cement pastes [61]. For the quantification of the disjoining pressure effect on the cement paste deformation, its influence area as a function of RH should be known. In hardened cements, this can be done through the water vapor sorption isotherms. However, in hardening pastes, this quantification is difficult, due to limitations arising from the rapidly changing properties of cement as hydration proceeds, but also the weak structure of the early-stage cement that is prone to drying shrinkage cracking [31]. While several investigations have tried to interpret the effect of disjoining pressure in cementitious materials [62, p. 19], [63–67], a reliable quantification of disjoining pressure and its role in predicting the shrinkage of cementitious materials is still unavailable.

3.8 *Typical Standards/Experiments to Measure Self-desiccation*

Internal Relative Humidity Measurement

The evolution of self-desiccation is strongly linked to reductions in internal RH, as previously discussed. The RH levels are therefore a strong indicator for self-desiccation. In concrete, RH is typically measured with commercially available hygrometers. Several types of RH sensors can be used for this purpose, namely, chilled mirror, resistive, and capacitive. However, these sensors show some drawbacks related to cost, accuracy at high RH levels and direct exposure to fresh concrete. To overcome these issues, Grasley et al. proposed a system consisting of small commercial sensors in plastic Gore-Tex caps which can be embedded in fresh concrete [26]. The system allows vapor, but not water, to be transmitted, reaching equal vapor pressure within the surrounding air of the sensor and concrete pores.

Various standards exist for monitoring internal RH in concrete. ASTM F2170 is a standard used to measure the moisture content of concrete slabs using in-situ probes. Moisture from concrete floor slabs can impact the performance of flooring systems like resilient and textile coverings or coatings. To prevent issues, manufacturers typically require moisture testing before installation, with internal relative humidity testing being one method. If excessive moisture remains after installation, it can lead to failures such as debonding, deterioration of finishes, and microbial growth.

RH in concrete can also be determined through the Nordtest Method NT Build 492. The method involves measurement on the internal RH on concrete pieces at age 1 month, providing a standard quality control procedure on self-desiccation, an important information in the construction industry.

Hygroscopic sorption isotherms are another procedure for determining the equilibrium moisture content in cementitious materials at various RH levels (ASTM C1498-04a 2016). The method can provide the adsorption isotherm (sorption isotherm describing the wetting process of the material from the oven-dry condition) and the desorption isotherm (sorption isotherm describing the drying process of a material from the state of absolute saturation with water).

Autogenous Shrinkage Measurement

Apart from internal RH, autogenous shrinkage is measured using non-contact displacement sensors, LVDTs, or strain gauges attached on opposite sides of the specimens to track volume reduction due to self-desiccation. Typical standards that exist are ASTM C1698-09, EN 12,390-16:2019, JCI (1998). All the above involve the measurement of autogenous strain (shrinkage or expansion) in cement paste and mortar without external drying effects, except for EN 12,390-16 which also covers drying shrinkage. ASTM C1698-09 is regarded as the most suitable one for monitoring hardening cement paste or mortar. Unlike ASTM C1698-09, JSI (1998) applies to larger concrete specimens, making it more representative of practical applications. A summary of the standards used for autogenous shrinkage measurement is shown in Table 1.

Apart from typical standards, RILEM recommendations provide detailed test methods for autogenous shrinkage when internal curing is used to mitigate self-desiccation (RILEM TC 196-ICC, RILEM TC 260-RSC). Various researchers have also proposed their own methods for determining autogenous shrinkage [68–74].

Pore Structure and Water Absorption Testing

Methods for characterizing the pore structure are also important, considering the crucial role of the latter in the self-desiccation process. These methods assess the water retention capacity and porosity, helping to evaluate the impact of self-desiccation on concrete. Among them are:

- ASTM C642-21, which is useful in developing the data required for conversions between mass and volume for concrete (density, percent absorption, and percent voids in hardened concrete).
- ASTM C1202 which provides the concrete's resistance to chloride penetration and is closely linked to permeability.

Table 1 Comparison of various standards for measuring autogenous shrinkage in cementitious systems

Standard	Specimen type	Moisture control	Measurement method	Preferred for
ASTM C1698-09	420 × 29 mm corrugated tubes of cement or mortar	Fully sealed	Dilatometer	Early-age autogenous shrinkage
EN 12,390-16:2019	Prisms or cylinders	Sealed (autogenous), exposed (drying)	Any length measuring device not relying on friction contact	Standardized shrinkage testing in hardened concrete (1 day onwards)
JCI (1998)	400 × 100 × 100 mm prisms	Sealed with tape or film	LVDT on a measuring frame	Full-scale concrete shrinkage testing

- MIP (Mercury Intrusion Porosimetry) and BET (Brunauer–Emmett–Teller) [58] techniques are used to characterize the pore size, volume, and porosity of materials, next to other techniques such as Dubinin-Astakhov or Dubinin-Radushkevich methods [75], amongst others.

Although not directly linked to self-desiccation, all aforementioned methods can provide a good indication on the evolution of the latter [76].

3.9 *Using SAP for Mitigating Self-desiccation*

As aforementioned, self-desiccation is a result of chemical shrinkage taking place in the absence of additional water. As self-desiccation occurs, capillary pressure increases and internal RH levels will decrease as cement hydrates, as indicated by the Kelvin Eq. (3). In low w/c mixtures, the drop in internal RH levels requires additional water to be provided to mitigate self-desiccation shrinkage. In this sense, internal curing mechanisms such as pre-saturated lightweight aggregates (LWA) or superabsorbent polymers (SAPs) have been proven very effective. The purpose of internal curing is to provide readily available additional water to keep the capillary porosity of the hydrating cement saturated, minimizing, therefore, autogenous stresses and strains [6]. In turn, the additional water will result in an increase of the degree of hydration and hence increased strength and reduced transport coefficients. Conventionally, this additional water was provided through external curing techniques such as ponding, misting, fogging or covering of the concrete surface. The advancement of concrete technology led to the use of high-strength, low w/c ratio concretes, which are characterized by very low capillary porosity [38], preventing external curing water from penetrating deep into the matrix. Therefore, the surface remains saturated, while the bulk volume undergoes significant self-desiccation [6].

Internal curing can provide the necessary water quantity needed throughout the self-desiccation process, with the proper rate (when exposed to ambient environmental conditions) and a proper homogeneity within the bulk volume, resulting in a well-saturated and stress-free cementitious matrix. In this manner, early-age cracking is minimized.

Even though LWA can successfully provide internal curing, there are certain drawbacks connected to the technique, such as difficulties controlling consistency and a significant strength and stiffness reduction [3]. On the contrary, SAP inclusion is more straightforward and shows less drawbacks. SAPs are added in concrete in a dry state, as opposed to the pre-saturated LWA. Moreover, SAPs permit free design of the pore shape and size during concrete hardening [3]. Additionally, the negative impact on the mechanical properties is reduced. For the aforementioned reasons, SAPs have been widely used for mitigating autogenous strains and self-desiccation [77–82], next to plastic shrinkage [79, 83].

The inclusion of SAPs in concrete and their impact on autogenous shrinkage mitigation were studied by Jensen and Hansen [84]. A 0.6 m% of SAP addition can maintain internal RH levels and successfully mitigate autogenous shrinkage in cement pastes. The latter behaviour has been widely studied in literature [24, 77, 78, 81, 85, 86]. An interlaboratory study of 13 institutes investigated the autogenous shrinkage evolution of HPC (w/c of 0.3 and 10% per cement weight of silica fume inclusion) enhanced with 0.3 m% SAPs, showing an autogenous shrinkage mitigation of 109%, 84% and 73% during the first three days, respectively, and 58%, 52% and 42% at later ages of 7, 14 and 28 days, respectively [87].

The timely delivery of water by the SAPs, although beneficial, is highly dependent on the SAP type and size, as well as the environmental conditions to which concrete is exposed. SAPs can be categorized as either self-releasing or retentive [81, 88]. The former releases the entrained water without any external trigger after casting, or due to vibration or compaction and can lead to an increase in the effective w/c ratio, increasing plastic shrinkage and the associated cracking risk. The latter retains the water until it can successfully be used for internal curing during the self-desiccation process, mitigating therefore, autogenous shrinkage. The releasing behaviour is based on the anionic functionalities present in an SAP's internal polymer network, which interact with cations present in the cementitious pore fluids. These interactions, particularly with K^+, Na^+, and $Ca2^+$, accelerate fluid desorption and lower their absorption capacity [77, 89]. Even though all SAPs can mitigate autogenous shrinkage to some extent, retentive SAPs continue to mitigate shrinkage for longer times (up to 7 days) compared to self-releasing SAPs [81]

Another important factor influencing SAP desorption is the environmental conditions. Shortly after concrete mixing, water flows from the cement matrix to the SAPs due to the higher ion concentration within the SAP particles. As hydration progresses and ions accumulate in the matrix's pore solution, this flow reverses, causing the SAPs to release water back into the matrix [90]. This occurs through increased surface tension and capillary pressure, as well as alterations in the ion concentration kinetics, which directly impact the SAP desorption rate. In unsealed systems, evaporation leads to premature depletion of free water, causing SAPs to release their stored water more quickly than in sealed systems. Elevated temperatures and windy conditions have the same effect, increasing the risk of early-age shrinkage and resulting cracking. For the aforementioned reasons, it is essential to detect and control SAP desorption to ensure optimal internal curing performance. Controlled SAP desorption can optimise curing and help maintain SAP pores saturated for longer, reducing thermal deformation and therefore cracking [91]. It is not only beneficial for preventing early internal curing water exhaustion, but also for optimizing curing without compromising mechanical properties. The amount of added SAP (and thus internal curing water) directly impacts compressive strength through increased (macro)porosity [80]. Lower dosages of SAP will have a less detrimental effect on compressive strength; however, they will be less effective in mitigating shrinkage. In this context, active control enhances SAP performance over a longer period [92], even allowing for a reduced SAP dosage, thus extending internal curing and offering prolonged protection during the sensitive early stages of material development.

4 Carbonation Shrinkage

Carbonation shrinkage is the chemical volume reduction due to the formation from $Ca(OH)_2$ to $CaCO_3$ in concrete in the presence of atmospheric CO_2. $Ca(OH)_2$, also known as Portlandite, forms abundantly in the concrete matrix as hexagonal plate-like crystals. If pozzolanic supplementary cementitious materials are used to partly replace cement, the Portlandite is consumed by a pozzolanic reaction, leaving less available for the long-term carbonation shrinkage [93].

This mechanism is often referred to just as carbonation, but it is always associated with a shrinkage phenomenon, even though it is often just the outer few millimetres of good, durable concrete. The negative impact of carbonation is mostly not the associated shrinkage, but more importantly, the lowering of the pH of concrete from the exposed surface. If this carbonation front reaches the reinforcing steel, the passivation layer around the steel is broken down and the steel in the reinforcing is prone to start corroding and negatively impacting the durability of the structure [94].

It is, however, difficult to isolate drying shrinkage and carbonation shrinkage from each other as shrinkage is not typically tested in a CO_2 free environment. The carbonation shrinkage has been found to be at its maximum at relative humidities between 50 and 60% [95]. At low relative humidities there is a lack of water which is required for the carbonation reaction to take place while at high humidities, the pores are typically filled with water which makes it difficult for the CO_2 to penetrate in the concrete.

Carbonation shrinkage is explained briefly in this report, but according to the knowledge of the authors no active control is done on carbonation (shrinkage) of concrete, other than maybe physical or chemical repair after some time. Physical repair would include removing the carbonated layer of concrete and chemical repair is when the carbonated concrete is treated chemically.

5 Plastic Shrinkage and Associated Cracking

Plastic shrinkage is in essence the shrinkage of concrete in the fresh state while being exposed to the environment. The evaporation rate is typically more than just conventional lab exposed conditions as adding a movement of air (wind) is critical in obtaining enough evaporation for plastic shrinkage cracking to occur.

Plastic shrinkage and drying shrinkage are similar with one aspect defining the difference between the two. Plastic shrinkage is caused by the drying of concrete still in the fresh/semi liquid state while the drying shrinkage is caused by the drying of hardened concrete. The typical measure of the transition point between plastic and drying shrinkage is the point of the Final Setting time which can easily be determined through a number of ways (e.g. ASTM C403) The mechanism of both shrinkage types is the forming of receding menisci of water in pore which causes a suction/negative pressure which results in shrinkage of the concrete. This mechanism is the same as explained in Sect. 3.1 for self-desiccation, namely capillary action, however the way of water loss is different. For self-desiccation the water is lost to the cement hydration reaction while for plastic shrinkage (and drying shrinkage for that matter) the water is lost to the environment through evaporation.

Plastic shrinkage has been shown to start occurring at the time step where all the bleed water has evaporated and further evaporation results in the reduction of water from the concrete body itself [96, 97]. This is as water has to evaporate from the concrete before capillary action can start [98]. It is important to note, for the purpose of active control of concrete, that this capillary action can be measured in fresh concrete by using tensiometers [98–101]. It is shown later in this report (chapter "Active Control of Cementitious Slabs and Roads") that these measurements can be used to actively control fresh concrete to reduce plastic shrinkage.

Typical negative capillary pressures developed in fresh concrete, known in unsaturated soil mechanics as suction, are shown in Fig. 5 [100]. The sensors first used were only capable of reading a negatively capillary pressure of at the most -100 kPa. Sensors have since been improved and can now read to the order of -1800 kPa, which is roughly around 18 times negative atmospheric pressure. Work is ongoing in improving these sensors, which is often still held back by a 24-h active pre-saturation phase and high costs as they become sacrificial after concrete has reached the Final Setting Time.

It has been shown that negative capillary pressure can be reversed almost immediately by adding water to the exposed surface of the concrete [98, 101]. These sensors can pick up this instant change and monitor the increase of negative pressure again. This makes it possible for this shrinkage mechanism to be used to control the internal

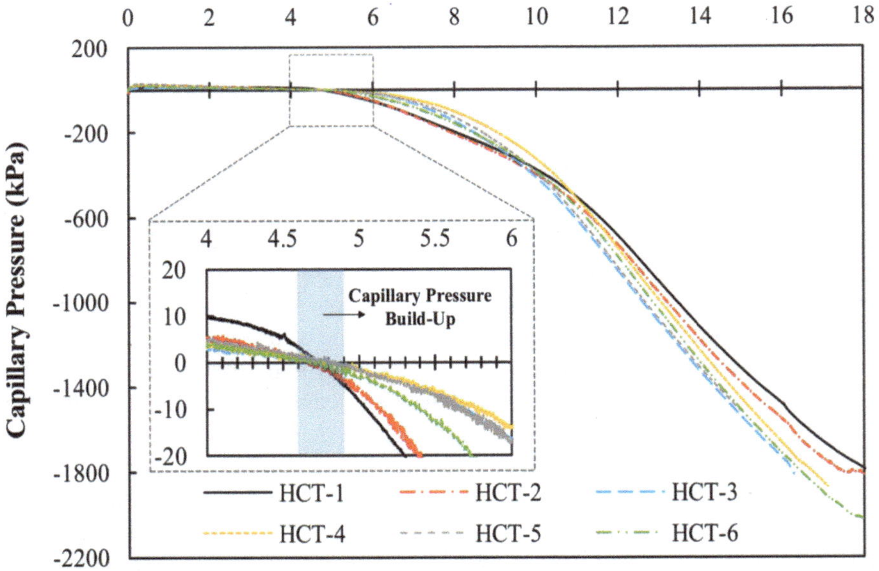

Fig. 5 Typical capillary pressure readings on concrete using high capacity tensiometers [100]

capillary pressure in concrete to improve the performance (cracking resistance) of fresh concrete when exposed to medium to extreme evaporation rates.

The magnitude of plastic shrinkage depends strongly on the evaporation rate, rate and volume of bleed water and on the w/c ratio, the porosity of the cementitious matrix, but also the presence of mineral admixtures and micro fibres [102–105]. Higher temperatures, low humidity and strong wind speeds can accelerate evaporation, and therefore, plastic shrinkage [106].

If concrete is fresh and still before initial set, plastic shrinkage typically does not cause cracking as the concrete material and associated strains can be redistributed even with restraint. Therefore, if cracks are seen before the initial set, they are typically due to settlement [107] as these tensile strains are significantly larger in magnitude. Plastic shrinkage cracking typically occurs between the initial and final set of concrete after which, any further crack development or increase of width is due to drying shrinkage as per definition. It has been shown that the typical plastic shrinkage crack growth accelerates after initial setting time and the crack width stabilises after final setting time [97].

This early-age plastic shrinkage will only lead to cracking if the concrete is restrained, either by reinforcing, formwork shapes, friction on formwork or substrates and lastly, internal restraint by means of a plastic shrinkage gradient. Plastic shrinkage cracking is thus related not only to the magnitude of plastic shrinkage, but also the boundary conditions. Work has also been done on defining the interaction between plastic shrinkage cracking and plastic settlement cracking as they appear more or less in the same time period (within 24 h) [107]. Many factors influence the probability and magnitude of plastic shrinkage cracking, including environmental factors (as mentioned before), the mixture proportions, the size of the concrete member [108, 109].

The tensile stresses created by the capillary pressure are parallel to one another and have typically an angle of 45°, with respect to the casting direction [102]. However, plastic shrinkage cracks can also form randomly [110], creating a map-like pattern. Factors influencing the pattern of plastic cracks involve the surface finishing direction and tamping marks [111, 112]. These cracks can have detrimental effects on the structure's durability as the cracks become pathways for water and other deleterious materials to ingress [113, 114].

6 Drying Shrinkage

By the time of final set a significant structure of pores has developed in the concrete, the concrete is now in the hardened state. If the water movement is permitted between the concrete pores and the environment, drying occurs with the water menisci in the pores and causes a contraction. This contraction will continue for a number of months after the final setting time, depending on the environmental conditions and the thickness of the concrete.

Drying shrinkage, already mentioned in the previous section, is defined as the shrinkage of concrete in the hardened state due to the loss of moisture to the environment, thus the loss of internal moisture within the porous structure, which seeks to achieve hygral equilibrium with the relative humidity (RH) of the surrounding environment [115]. The biggest difference in the mechanism of plastic shrinkage and drying shrinkage is the state of the concrete (fresh versus hardened) and the rate at which the drying and resulting shrinkage happens. Drying shrinkage occurs at a much slower rate than plastic shrinkage. Both, if movement is restrained, can cause cracking.

Environmental conditions, such as high temperature, strong wind, low RH, will induce drying conditions and will, therefore, accelerate the drying shrinkage evolution. Drying shrinkage is also strongly influenced by the curing method, restraints (mould and/or reinforcement), pore size distribution, the thermodynamic behaviour of water within the pores, and the morphology of hydrates in the microstructure [116]. Volumetric instability is primarily driven by four mechanism-based theories: changes in surface free energy (Gibbs-Bangham effect), capillary tension, disjoining pressure, and the movement of interlayer water in the C-S-H gel [117] while it is believed that only capillary tension is responsible for plastic shrinkage. The Gibbs-Bangham mechanism is a primary contributor to drying shrinkage stress, mainly functioning at RH levels below 40%. Meanwhile, disjoining pressure and capillary tension work together in the RH range from 40 to 95% [20]. At the critical RH range of 40–50%, the dominant mechanism shifts from surface free energy to capillary pressure, with capillary pressure having its greatest impact on volumetric deformation [116]. To address these issues, proper curing is essential to maintain adequate internal RH levels and prevent drying shrinkage cracking in the sensitive early-stage [118] but also prevent significant volume reductions [2].

Drying shrinkage is mentioned briefly in this chapter to complete the list of shrinkage mechanisms and phenomena in concrete. Drying shrinkage will not be addressed later in the report as there is, to the authors' knowledge, currently no work done on the active control of drying shrinkage.

7 Concluding Summary

This chapter on shrinkage and associated cracking looked into the mechanisms and phenomena of shrinkage in concrete, the causes and associated crack formation. The main mechanisms/phenomena addressed are chemical shrinkage, self-desiccation, carbonation shrinkage, plastic shrinkage, and drying shrinkage. This chapter also includes a preliminary overview of the properties that may be actively controlled. Further details on the active control options, and a summary of the already published research results in this context can be found in chapter "Active Control of Cementitious Slabs and Roads".

References

1. Holt, E.: Contribution of mixture design to chemical and autogenous shrinkage of concrete at early ages. Cem. Concr. Res. **35**(3), 464–472 (2005)
2. Holt, E., Leivo, M.: Cracking risks associated with early age shrinkage. Cement Concr. Compos. **26**(5), 521–530 (2004)
3. Jensen, O.M., Hansen, P.F.: Water-entrained cement-based materials I. Principles and theoretical background. Cem. Concr. Res. **31**(4) (2001)
4. Ghanem, H., Ramadan, R., Khatib, J., Elkordi, A.: A review on chemical and autogenous shrinkage of cementitious systems. Materials **17**(2), 283 (2024)
5. Geiker, M., Knudsen, T.: Chemical shrinkage of Portland cement pastes. Cem. Concr. Res. **12**(5), 603–610 (1982)
6. Bentz, D.P., Weiss, W.J.: Internal Curing: A 2010 State-of-the-Art Review (2011)
7. Le Chatelier, H.: Sur les Changements de Volume Qui Accompagnent le Durcissement des Ciments. Bull. Société L'encouragement Pour L'industrie Natl. **5**, 54–57 (1900)
8. Yodsudjai, W., Wang, K.: Chemical shrinkage behaviour of pastes made with different types of cements. Constr. Build. Mater. **40**, 854–862 (2013)
9. Beltzung, F., Wittmann, F.H., Holzer, L.: Influence of composition of pore solution on drying shrinkage. In: Presented at the Creep, Shrinkage and Durability Mechanics of Concrete and Other Quasi-Brittle Materials, pp. 39–48. MIT (2001)
10. Bentz, D.P., Sant, G., Weiss, J.: Early-age properties of cement-based materials. I: influence of cement fineness. J. Mater. Civ. Eng. **20**(7), 502–508 (2008)
11. Wu, L., Farzadnia, N., Shi, C., Zhang, Z., Wang, H.: Autogenous shrinkage of high performance concrete: a review. Constr. Build. Mater. **149**, 62–75 (2017)
12. Persson, B.: Experimental studies on shrinkage of high-performance concrete. Cem. Concr. Res. **28**(7), 1023–1036 (1998)
13. Lura, P., Couch, J., Jensen, O.M., Weiss, J.: Early-age acoustic emission measurements in hydrating cement paste: evidence for cavitation during solidification due to self-desiccation. Cem. Concr. Res. **39**(10), 861–867 (2009)
14. Fisher, L.R., Israelachvili, J.N.: Experimental studies on the applicability of the Kelvin equation to highly curved concave menisci. J. Colloid Interface Sci. **80**(2), 528–541 (1981)
15. Young, T.: III. An essay on the cohesion of fluids. Philos. Trans. R. Soc. Lond. **95**, 65–87 (1805)
16. Wittmann, F.H.: Interaction of hardened cement paste and water. J. Am. Ceram. Soc. **56**(8), 409–415 (1973)
17. Wittmann, F.H.: On the action of capillary pressure in fresh concrete. Cem. Concr. Res. **6**, 49–56 (1976)
18. Badmann, R., Stockhausen, N., Setzer, M.J.: The statistical thickness and the chemical potential of adsorbed water films. J. Colloid Interface Sci. **82**(2), 534–542 (1981)
19. Adamson, A.W., Gast, A.P.: Physical chemistry of surfaces, vol. 150. Interscience Publishers, New-York (1967)
20. Ye, H., Radlińska, A.: A review and comparative study of existing shrinkage prediction models for portland and non-portland cementitious materials. Adv. Mater. Sci. Eng. **1**, 2418219 (2016)
21. Radlińska, A., Rajabipour, F., Bucher, C., Henkensiefken, R., Sant, G., Weiss, J.: Shrinkage mitigation strategies in cementi-tious systems: a closer look at differences in sealed and unsealedbehaviour. Transp. Res. Rec. **2070**, 59–67 (2008)
22. Bentz, D.P.: Curing with shrinkage-reducing admixtures. Concr. Int. **27**, p55 (2005)
23. Meddah, M.S., Tagnit-Hamou, A.: Pore structure of concrete with mineral admixtures and its effect on self-desiccation shrinkage. ACI Mater. J. **106**(3), 241 (2009)
24. Snoeck, D., Jensen, O.M., De Belie, N.: The influence of superabsorbent polymers on the autogenous shrinkage properties of cement pastes with supple-mentary cementitious materials. Cem. Concr. Res. **74**, 59–67 (2015)
25. Li, Y., Bao, J., Guo, Y.: The relationship between autogenous shrinkage and pore structure of cement paste with mineral admixtures. Constr. Build. Mater. **24**(10), 1855–1860 (2010)

26. Grasley, Z.C., Lange, D.A., D'Ambrosia, M.D., Villalobos-Chapa, S.: Rela-tive humidity in concrete. Concr. Int. **28**(10), 51–57 (2006)
27. Kim, J.-K., Lee, C.-S.: Moisture diffusion of concrete considering self-desiccation at early ages. Cem. Concr. Res. **29**(12), 1921–1927 (1999)
28. El-Dieb, A.: Self-curing concrete: water retention, hydration and moisture transport. Constr. Build. Mater. **21**(6), 1282–1287 (2007)
29. Grasley, Z.C., Lange, D.A., D'Ambrosia, M.D.: Internal relative humidity and drying stress gradients in concrete. Mater. Struct. **39**(9), 901–909 (2006)
30. Grasley, Z.C., Lange, D.A.: Thermal dilation and internal relative humidity of hardened cement paste. Mater. Struct. **40**(3), 311–317 (2007)
31. Lura, P., Jensen, O.M., van Breugel, K.: Autogenous shrinkage in high-performance cement paste: an evaluation of basic mechanisms. Cem. Concr. Res. **33**(2), 223–232 (2003)
32. Jensen, O.M.: Autogenous deformation and RH-change—self-desiccation and self-desiccation shrinkage. Technical University of Denmark, Lyngby (1993)
33. Jiang, Z., Sun, Z., Wang, P.: Study on autogenous relative humidity change in high performance concrete. J. Chin. Ceram. Soc. **31**(8), 770–773 (2003)
34. Snoeck, D., Pel, L., De Belie, N.: The water kinetics of superabsorbent poly-mers during cement hydration and internal curing visualized and studied by NMR. Sci. Rep. **7**(9514) (2017). https://doi.org/10.1038/s41598-017-10306-0
35. Yang, Q.: Inner relative humidity and degree of saturation in highperformance concrete stored in water or salt solution for 2 years. Cem. Concr. Res. **29**(1), 45–53 (1999)
36. Yang, Q.B., Zhang, S.Q.: Self-desiccation mechanism of high-performance concrete. J. Zhejiang Univ. Sci. A **5**(12), 1517–1523 (2004)
37. Wyrzykowski, M., Lura, P.: Effect of relative humidity decrease due to self-desiccation on the hydration kinetics of cement. Cem. Concr. Res. **85**, 75–81 (2016)
38. Powers, T.C.: A discussion of cement hydration in relation to the curing of concrete. Highw. Res. Board Proc. **27**, 178–188 (1947)
39. Jensen, O.M.: Thermodynamic limitation of self-desiccation. Cem. Concr. Res. **25**, 157–164 (1995)
40. Jensen, O.M., Hansen, P.F., Lachowski, E.E., Glasser, F.P.: Clinker mineral hydration at reduced relative humidities. Cem. Concr. Res. **29**, 1505–1512 (1999)
41. Flatt, R.J., Scherer, G.W., Bullard, J.W.: Why alite stops hydrating below 80% relative humidity. Cem. Concr. Res. **41**, 987–992 (2011)
42. Lothenbach, B., Scrivener, K.L., Hooton, D.: Supplementary cementitious materials. Cem. Concr. Res. **41**(12), 1244–1256 (2011)
43. Damidot, D., Lothenbach, B., Herfort, D., Glasser, F.P.: Thermodynamics and cement science. Cem. Concr. Res. **41**(7), 679–695 (2011)
44. Zhang, M., Tam, C., Leow, M.: Effect of water-to-cementitious materials ratio and silica fume on the autogenous shrinkage of concrete. Cem. Concr. Res. **33**(10), 1687–1694 (2003)
45. Tian, Q.: Study on shrinkage and mechanism of cement-based material at low water to binder ratio incorporating high volume mineral admixtures. Southeast University (2006)
46. Scherer, G.W.: Theory of drying. J. Am. Ceram. Soc. **73**(1), 3–14 (1990)
47. Scherer, G.W.: Drying, shrinkage, and cracking of cementitious materials. Transp. Porous Media **110**(2), 311–331 (2015)
48. Tazawa, E., Miyazawa, S.: Influence of cement composition on autogenous shrinkage of concrete. In: 10th International Conference on Cement Chemistry, Goteborg (1997)
49. Tazawa, E., Miyazawa, S.: Influence of cement and admixture on autoge-nous shrinkage of cement paste. Cem. Concr. Res. **25**(2), 281–287 (1995)
50. Tazawa, E., Miyazawa, S.: Influence of constituents and composition on autogenous shrinkage of cementitious materials. Mag. Concr. Res. **49**(178), 15–22 (1997)
51. van Breugel, K., van Tuan, N.: Autogenous shrinkage of HPC and ways to mitigate it. Key Eng. Mater. **629–630**, 3–20 (2014)
52. Bullard, J.W., et al.: Mechanisms of cement hydration. Cem. Concr. Res. **41**(12), 1208–1223 (2011)

53. Bentz, D.P., Garboczi, E.J., Haecker, C.J., Jensen, O.M.: Effects of cement particle size distribution on performance properties of Portland cement-based materials. Cem. Concr. Res. **29**(10), 1663–1671 (1999)
54. Bentz, D.P., Jensen, O.M., Hansen, P.F., Olesen, J.F., Stang, H., Haecker, C.J.: Influence of cement particle-size distribution on early age autogenous strains and stresses in cement-based materials. J. Am. Ceram. Soc. **84**(1), 129135 (2001)
55. Bentz, D.P., Haecker, C.J.: An argument for using coarse cements in highperformance concretes. Cem. Concr. Res. **29**(4), 615–618 (1999)
56. Turcry, P., Loukili, A., Barcelo, L., Casabonne, J.M.: Can the maturity con-cept be used to separate the autogenous shrinkage and thermal deformation of a cement paste at early age? Cem. Concr. Res. **32**(9), 1443–1450 (2002)
57. Ferraris, C., Wittmann, F.H.: Shrinkage mechanisms of hardened cement paste. Cem. Concr. Res. **17**, 453–464 (1987)
58. Hagymassy, J., Brunauer, S., Mikhail, R.S: Pore structure analysis by water vapor adsorption—I. t-curves for water vapor. J. Colloid Interface Sci. **29**(3) (1969)
59. Neimark, A.V., Ravikovitch, P.I., Vishnyakov, A.: Bridgingscales from molecular simulations to classical thermodynamics: density functional theory of capillary condensation innanopores. J. Phys.: Condens. Matter **15**(3), 347–365 (2003)
60. Bazant, Z.P., Bazant, M.Z.: Theory of sorption hysteresis innanoporous solids: part I: snap-through instabilities. J. Mech. Phys. Solids **60**(9), 1644–1659 (2012)
61. Nielsen, L.F.: A research note on sorption, pore size distribution, and shrinkage of porous materials. Building Materials Laboratory, The Technical University of Denmark, Lyngby, TR 245/91 (1991)
62. Powers, T.C.: The thermodynamics of volume change and creep. Materiaux et Constr. **1**(6), 487–507 (1968)
63. Feldman, R.F., Sereda, P.J.: A new model for hydrated portland cement and its practical implications. Eng. J. **53**(8–9) (1970)
64. Bazant, Z.P.: Thermodynamics of interacting continua with surfaces and creep analysis of concrete structures. Nucl. Eng. Des. **20**(2), 477–505 (1972)
65. Bazant, Z.P.: Mathematical model for creep and thermal shrinkage of concrete at high temperature. Nucl. Eng. Des. **76**(2), 183–191 (1983)
66. Beltzung, F., Wittmann, F.H.: Role of disjoining pressure in cement based materials. Cem. Concr. Res. **35**(12), 2364–2370 (2005)
67. Ye, H.: Creep mechanisms of calcium–silicate–hydrate: an overview of recent advances and challenges. Int. J. Concr. Struct. Mater. **9**(4), 453–462 (2015)
68. Amin, M.N., Kim, J.S., Dat, T.T., Kim, J.K.: Improving test methods to measure early age autogenous shrinkage in concrete based on air cooling. IES J. Part A: Civ. Struct. Eng. **3**(4), 244–256 (2010)
69. Jensen, O.M., Hansen, P.F.: A dilatometer for measuring autogenous de-formation in hardening Portland cement paste. Mater. Struct. **28**, 406–409 (1995)
70. Radocea, A.: Autogenous volume change of concrete at early ages model and experimental data. Lund University, Lund Institute of Technology, Lund, TVBM-3075 (1997)
71. Hammer, T.A.: Test methods for linear measurement of autogenous shrinkage before setting. In: Proceedings of International Workshop on Autogenous Shrink age of concrete, pp 143–154. E & FN Spon, Hiroshima (1998)
72. Bjøntegaard, O.: Thermal dilation and autogenous deformation as driving forces to self-induced stresses in high performance concrete. PhD dissertation, The Norwegian University of Science and Technology, Norway (1999)
73. Holt, E.E.: Early age autogenous shrinkage of concrete. PhD dissertation, University of Washington in Seattle, Seattle (2001)
74. Ostergaard, T., Jensen, O.M.: A thermal comparator sensor for measuring autogenous deformation in hardening Portland cement paste. Mater. Struct. **36**, 661–665 (2003)
75. Dubinin, M.M.: Physical adsorption of gases and vapors in micropores. In: Danielli, J.F., Rosenberg, M.D., Cadenhead, D.A. (eds.) Progress in Surface and Membrane Science, vol. 9, pp. 1–70. Academic Press (1975)

76. Snoeck, D., et al.: The influence of different drying techniques on the water sorp-tion properties of cement-based materials. Cem. Concr. Res. **64**, 54–62 (2014)
77. Schröfl, C., Erk, K.A., Siriwatwechakul, W., Wyrzykowski, M., Snoeck, D.: Recent progress in superabsorbent polymers for concrete. Cem. Concr. Res. **151**(106648) (2022)
78. Mechtcherine, V., et al.: Application of super absorbent polymers (SAP) in con-crete construction-update of RILEM state-of-the-art report. Mater. Struct. **54**(80) (2021)
79. Boshoff, W.P., et al.: The effect of superabsorbent polymers on the mitigation of plastic shrinkage cracking of conventional concrete, results of an inter-laboratory test by RILEM TC 260-RSC. Mater. Struct. **53**(79) (2020)
80. Hasholt, M.T., Jensen, O.M., Kovler, K., Zhutovsky, S.: Can superabsorbent polymers mitigate autogenous shrinkage of internally cured concrete without compromising the strength? Constr. Build. Mater. **31**, 226–230 (2012)
81. Schröfl, C., Mechtcherine, V., Gorges, M.: Relation between the molecular structure and the efficiency of superabsorbent polymers (SAP) as concrete admix-ture to mitigate autogenous shrinkage. Cem. Concr. Res. **42**(6) (2012). https://doi.org/10.1016/j.cemconres.2012.03.011
82. Assmann, A., Reinhardt, H.W.: Tensile creep and shrinkage of SAP modified concrete. Cem. Concr. Res. **58**, 179–185 (2014)
83. Snoeck, D., Pel, L., De Belie, N.: Superabsorbent polymers to mitigate plastic drying shrinkage in a cement paste as studied by NMR. Cem. Concr. Compos. **93**, 54–62 (2018)
84. Jensen, O.M., Hansen, P.F.: Water-entrained cement-based materials II. Experimental observations. Cem. Concr. Res. **32**(6) (2002)
85. De Meyst, L., et al.: Parameter study of superabsorbent polymers for use in durable concrete structures. Materials **12**(9) (2019)
86. Yang, J., Sun, Z., De Belie, N., Snoeck, D.: Mitigating plastic shrinkage cracking in alkali-activated slag systems by internal curing with superabsorbent polymers. Cem. Concr. Compos. **134**(104784) (2022)
87. Mechtcherine, V., et al.: Effect of internal curing by using superabsorbent polymers (SAP) on autogenous shrinkage and other properties of a high-performance fine-grained concrete: results of a RILEM round-robin test, TC 225-SAP. Mater. Struct. **47**(3) (2014)
88. Siriwatwechakul, W., Siramanont, J., Vichit-Vadakan, W.: Behaviour of su-perabsorbent polymers in calcium- and sodium-rich solutions. J. Mater. Civ. Eng. **24**(8), 976–980 (2012)
89. Snoeck, D., Van Tittelboom, K., Steuperaert, S., Dubruel, P., De Belie, N.: Self-healing cementitious materials by the combination of microfibres and superabsorbent polymers. J. Intell. Mater. Syst. Struct. **25**(1)(1) (2014). https://doi.org/10.1177/1045389X12438623
90. Danish, A., Mosaberpanah, M.A., Salim, M.U.: Robust evaluation of super-absorbent polymers as an internal curing agent in cementitious composites. J. Mater. Sci. **56**, 136–172 (2021)
91. Wyrzykowski, M., Lura, P.: Controlling the coefficient of thermal expansion of cementitious materials—a new application for superabsorbent polymers. Cem. Concr. Compos. **35**(1), 49–58 (2013)
92. Korda, E., Tsangouri, E., Snoeck, D., De Schutter, G., Aggelis, D.G.: Active control of concrete curing monitored by acoustic emission. In: Proceedings of the RILEM Spring Convention and Conference 2024. RSCC 2024, Conference on advanced construction materials and processes for a carbon neutral society, pp. 412–420. Springer RILEM Publications S.A.R.L., Milano (2025)
93. von Greve-Dierfeld, S., et al.: Understanding the carbonation of concrete with supplementary cementitious materials: a critical review by RILEM TC 281-CCC. Mater. Struct. **53**(136) (2020)
94. Rozière, E., Loukili, A., Cussigh, F.: A performance based approach for durability of concrete exposed to carbonation. Constr. Build. Mater. **23**(1), 190–199 (2009)
95. Verbeck, G.J.: Carbonation of hydrated Portland cement. Res. Dept. Bull. **87** (1958)
96. Lerch, W.: Plastic shrinkage. ACI J. **53**, 797–802 (1957)
97. Combrinck, R., Boshoff, W.P.: Typical plastic shrinkage cracking behaviour of concrete. Mag. Concr. Res. **65**(8) (2013)

98. Slowik, V., Schmidt, M., Fritzsch, R.: Capillary pressure in fresh cement-based materials and identification of the air entry value. Cem. Concr. Compos. **30**, 557–565 (2008)
99. Slowik, V., Schmidt, M., Kässler, D., Eiserbeck, M.: Capillary pressure monitoring in plastic concrete for controlling early-age shrinkage cracking. J. Transp. Res. Board **2441**(1) (2014)
100. Jamali, A., Mendes, J., Nagaratnam, B., Lim, M.: A new four stage model of capillary pressure in early age concrete: Insights from high capacity tensiometers. Cem. Concr. Res. **161**(106955) (2022)
101. Deysel, R.C., Boshoff, W.P., Smit, M.S.: Implementing capillary pressure control measures to prevent plastic shrinkage cracking in concrete. Constr. Build. Mater. **397**(132407) (2023)
102. Slowik, V., Hübner, T., Schmidt, M., Villmann, B.: Simulation of capillary shrinkage cracking in cement-like materials. Cement Concr. Compos. **31**(7), 461–169 (2009)
103. Boshoff, W.P., Combrinck, R.: Modelling the severity of plastic shrinkage cracking in concrete. Cem. Concr. Res. **48**, 34–39 (2013)
104. Kayondo, M., Combrinck, R., Boshoff, W.P.: State-of-the-art review on plastic cracking of concrete. Constr. Build. Mater. **225**, 886–899 (2019)
105. Zhang, B., Zhu, H., Cheng, Y., Huseien, G.F., Shah, K.W.: Shrinkage mechanisms and shrinkage-mitigating strategies of alkali-activated slag compo-sites: a critical review. Constr. Build. Mater. **318**(125993) (2022)
106. Uno, P.J.: Plastic shrinkage cracking and evaporation formulae. Mater. J. **95**(4) (1998)
107. Combrinck, R., Steyl, L., Boshoff, W.P.: Interaction between settlement and shrinkage cracking in plastic concrete. Constr. Build. Mater. **185**(1–11) (2018)
108. Shaeles, C.A., Hover, K.C.: Influence of mix proportions and construc-tion operations on plastic shrinkage cracking in thin slabs. ACI Mater. J. **48**, 495–504 (1988)
109. Lura, P., Pease, B.J., Mazzotta, G.B., Rajabipour, F., Weiss, J.: Influence of shrinkage-reducing admixtures on development of plastic shrinkage cracks. ACI Mater. J. **104**(2), 187–194 (2007)
110. Cohen, M.D., Olek, J., Dolch, W.L.: Mechanism of plastic shrinkage cracking in portland cement and portland cement-silica fume paste and mortar. Cem. Concr. Res. **20**, 103–119 (1990)
111. Day, R., Clarke, J.: Plastic and thermal cracking. In: Advanced Concrete Technology, vol. 1, pp. 3–17. Butterworth Heinemann (2003)
112. Kurup, D.S., Mohan, M.K., Van Tittelboom, K., De Schutter, G., Santha-nam, M., Rahul, A.V.: Early-age shrinkage assessment of cementitious materials: a critical review. Cem. Concr. Compos. **145**(105343) (2024)
113. Qi, C.: Quantitative assessment of plastic shrinkage cracking and its effect on the corrosion of steel reinforcement. PhD dissertation, Purdue University, Indiana (2003)
114. Otieno, M.B., Alexander, M.G., Beushausen, H.-D.: Corrosion in cracked and uncracked concrete—influence of crack width, concrete quality and crack reopening. Mag. Concr. Res. **62**(6), 393–404 (2010)
115. Di Bella, C.: Drying shrinkage of cementitious materials at early age. PhD dissertation, ETH Zürich, Zürich (2016)
116. Wittmann, F.H.: Creep and shrinkage mechanisms. In: Creep and Shrinkage in Concrete Structures, pp 129–161. John Wiley & Sons Ltd (1982)
117. Kovler, K., Zhutovsky, S.: Overview and future trends of shrinkage re-search. Mater. Struct. **39**, 827–847 (2006)
118. Wang, J., Li, H., Tan, Y., Wang, J., Gao, Y.: Study on the shrinkage and restrained cracking of concrete with different surface curing methods. Mater. Struct. **58**(9) (2005)
119. Gowripalan, N.: Autogenous shrinkage of concrete at early ages. In: Presented at the ACMSM25. Lecture Notes in Civil Engineering. Springer, Singapore (2020)

Active Control of Concrete Pumping

Mert Yücel Yardımcı, Dimitri Feys, and Geert De Schutter

Abstract Pumping concrete is a very popular processing technique, applied every day on construction sites worldwide. Although the principles seem simple, daily practice can show many issues and challenges, including segregation and blocking. This chapter gives a brief introduction to the principles of concrete pumping and the key challenges faced. To overcome the challenges, the chapter discusses potential active control opportunities in concrete pumping.

Keywords Concrete · Pumping · Lubrication layer · Active control

1 Introduction

Concrete pumping is a common technique for delivering freshly mixed concrete to the placement location in modern construction. This method involves transporting fresh concrete through pipelines using hydraulic or mechanical pumps, and offers significant advantages, including increased efficiency, reduced labour, and easy delivery of fresh concrete to areas difficult to reach on the construction site. Compared to manually placing the fresh concrete mixture, delivering the fresh mixture by pumping is more efficient for construction sites where a continuous concrete supply is needed. Normally, a high variety of different concrete types can be pumped without facing a significant problem if the pumping operation is well-planned and the mixture is adequately designed. On the other hand, the pumping process can also be quite challenging and even can result in the blockage of the pipeline in case of mixtures with

M. Y. Yardımcı (✉)
Department of Civil Engineering, Dokuz Eylul University, Izmir, Türkiye
e-mail: mert.yardimci@deu.edu.tr

M. Y. Yardımcı · G. De Schutter
Magnel-Vandepitte Laboratory, Department of Structural Engineering and Building Materials, Ghent University, Ghent, Belgium

D. Feys
Missouri University of Science and Technology, Rolla, MO, USA

G. De Schutter and J. Sanjayan (eds.), *Active Control of Properties of Fresh and Hardening Cementitious Materials*, RILEM State-of-the-Art Reports 50,
https://doi.org/10.1007/978-3-032-30340-0_6

high slump-loss or insufficient cohesion. After starting the pumping operation, the only option to regulate the pumping process is to control the pumping pressure or discharge rate. Although the rheological properties, such as yield stress, viscosity and thixotropy of concrete have key roles in the success of pumping, there is currently no practical active control option of the rheological properties of concrete after the start of concrete pumping operations. This chapter gives a brief introduction to the principles of concrete pumping and the key challenges faced and discusses active control opportunities for concrete pumping.

2 Principles of Concrete Pumping

Concrete pumping relies on the use of pumps to push fresh concrete through a pipeline. This technique has been used since the 1960s [1] and it is assumed to be the most efficient and economical mode of transporting and placing fresh concrete into the desired place. The pumping process of the fresh concrete involves complex interactions between the rheological properties of the fresh concrete mixture, the pumping equipment, and the pipeline properties [2], which can influence the pumpability and final quality of the concrete [3]. Predicting the pumpability of concrete mixtures and selecting the appropriate pump with sufficient power is a difficult task. However, some concrete proportioning recommendations and design charts to predict concrete pressure are available [4]. Design charts, a simplified example of which is shown in Fig. 1, are used to estimate the pumping pressure required for pumping according to the desired flow rate, pipe diameter and length, and a relevant rheological parameter of the concrete (traditionally the slump value, but for modern binder types rather a viscosity parameter).

In such traditional design charts, as illustrated in the top part of Fig. 1, the pumpability of fresh concrete is evaluated considering the slump value of fresh concrete as the only parameter determining the flow behaviour of the concrete. Although the slump test can be assumed as a simple and good practical qualitative indicator for the yield stress of concrete [5], the slump value does not completely determine the rheology of fresh concrete, including the plastic viscosity of the mixture. Concrete mixtures with very different viscosities or aggregate ratios may have the same slump values but will have different resistance to flow during pumping. For example, according to such design charts, flowable concretes with high slump values require a lower pumping pressure. However, this is not always the case, as it has been shown that very flowable but generally more viscous concrete mixtures like self-compacting concrete (SCC) may even require higher pressure to be pumped compared to conventional vibrated concrete (CVC) [5, 6]. As a result, updated design charts have been developed considering a viscosity parameter instead of the slump value [8], as illustrated in the bottom part of Fig. 1. As shown in [9], the two main rheological parameters, i.e., yield stress and plastic viscosity, can be adversely affected by changes in the design parameters of concrete. Therefore, understanding the rheological insights behind the pumping behaviour of concrete is essential.

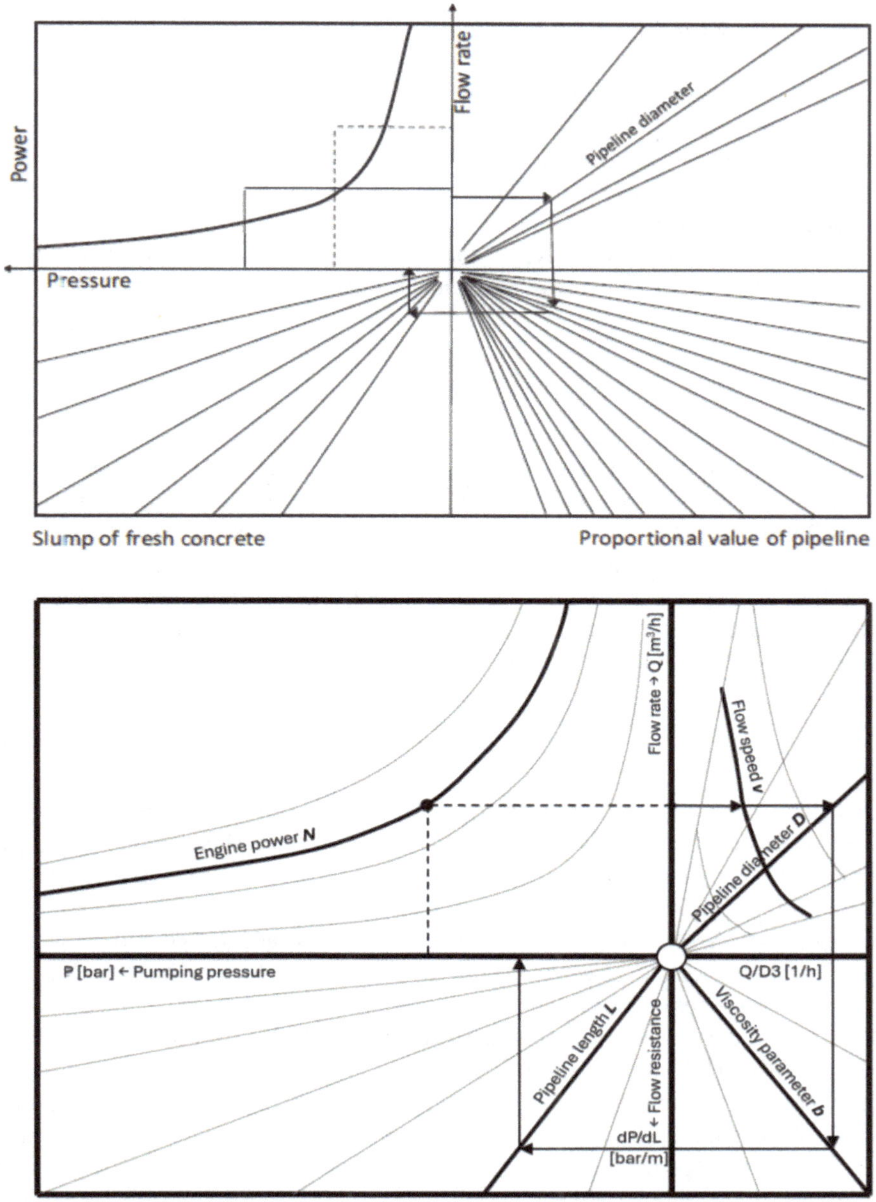

Fig. 1 Pumping pressure prediction chart example for conventional vibrated concrete [1] (top), for self-compacting concrete (bottom) (Adapted from [8])

2.1 Basics of Concrete Flow in a Pipeline

Concrete does not flow homogeneously inside the pipeline, instead, it flows with different flow profiles with two or three zones of different properties [10]. The reason why the concrete flowing inside the pipe has different flow profiles in the cross-section of the pipe is the existence of a yield stress in concrete, and the formation of a lubrication layer near the pipe surface.

When a flowing material is forced through a pipeline, the resulting shear stress (τ) induced by the pumping pressure (P) inside the cylindrical pipe follows Eq. (1).

$$\tau = \frac{r}{2}\frac{dP}{dL} \tag{1}$$

Where: r is the radial distance from the pipe center, dP/dL is the pressure gradient along the pipeline. Eq. (1) shows that the shear stress at the pipe wall ($r = R$, *radius of pipe*) reaches a maximum value τ, while it becomes zero at the pipe center ($r = 0$), as illustrated in Fig. 2a. The pressure gradient along the pipeline can roughly be assumed to be equal to the total pressure loss over a pipeline length ($dP/dL \approx \Delta P_{tot}/L$) as long as the pipes are straight with a constant diameter and the rheology does not change. Therefore, the shear stress occurring at the pipe wall (τ_w) can be expressed by Eq. (2) in terms of the pressure loss per unit length of the pipeline (Δp).

$$\tau_w = \frac{R}{2}\Delta p \tag{2}$$

This shear stress distribution given by Eq. (1) is a universal law, and for a fixed pressure loss the shear stress distribution and the maximum shear stress at the pipe wall, Eq. (2), are independent of the rheological properties of the pumping material. The corresponding shear rate can be calculated from the shear stress if the rheological parameters of the material are known. However, the shear rate and corresponding velocity profiles in the pipe will mainly depend on the competition between the evolving shear stress and the yield stress of the pumping material, as illustrated in Fig. 2 [11]. If the pumped material is a Newtonian material, which is a fluid that does not require a threshold yield stress to flow, an immediate flow will occur as soon as any shear stress is applied. In such a material, every adjacent layer of the material will be sheared from a zero value at the centre of the pipe, and a maximum shear rate will occur at the pipe wall (green line in Fig. 2b). This means that the material, under the assumption of no-slip at the pipe wall, will move with a parabolically increasing velocity profile from the centre of the pipe as illustrated in the green line in Fig. 2c. However, this is not the case if the pumped material has a certain yield stress to be overcome to initiate the flow (such as concrete). In non-Newtonian materials like cement paste and concrete, the adjacent layers of the material are not sheared (i.e., flow) in the regions where the shear stress induced by pumping pressure (τ) is smaller than the yield stress of the pumped material (τ_0). In such materials, a portion of the material along the central axis remains unsheared (yellow and purple lines in

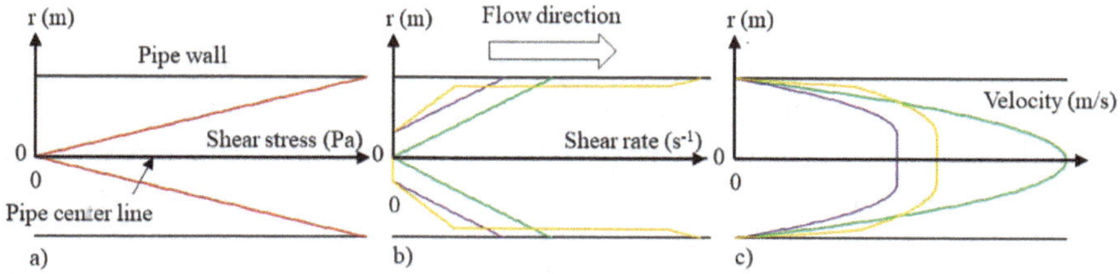

Fig. 2 Shear stress induced by pumping pressure **a**, evolving shear rate **b** and velocity **c** profiles in the pipe for Newtonian (green), and Bingham with lubrication layer (yellow). (Adopted from [11], solid black lines show the pipe walls)

Fig. 2b). This zone is called the plug zone and the uniform movement of concrete with a constant velocity is named plug flow (yellow and purple lines in Fig. 2c) [11].

The velocity profile can be obtained by integrating the shear rate profile [12] and the flow rate can then be obtained by integrating the velocity profile [11]. The shear rate of a Newtonian fluid through a cylindrical pipe under laminar flow conditions (Reynolds number Re < 2000) linearly evolves starting from a zero value at the pipe centre and reaching the highest value at the pipe wall. The shear stress is proportional to the velocity gradient, and the relation between the flow rate and total pressure loss can be expressed by Eq. (3) using the Hagen-Poiseuille relation [13, 14].

$$Q = \frac{\pi \Delta P_{tot} D^4}{128 \mu L} \tag{3}$$

Where Q is volumetric flow rate, ΔP_{tot} is the total pressure loss across the pipe length, μ is the dynamic viscosity of the fluid, L is the pipeline length, and D is the radius of the pipe. The Hagen-Poiseuille relation is valid for Newtonian materials (no yield stress, constant viscosity). On the other hand, mortar and concrete are non-Newtonian materials following the Bingham or the modified Bingham rheological models. Therefore, the shear rate distribution in Bingham materials against the shear stress distribution over the pipe cross-section is different from that of Newtonian materials, as the material has sheared regions ($\tau > \tau_0$) and unsheared regions ($\tau < \tau_0$). For Bingham materials like concrete, the relationship between pressure loss and flow rate is expressed by the Buckingham-Reiner equation, Eq. (4), as an extended form of the Hagen-Poiseuille relation [11, 15]. The extension of the Hagen-Poiseuille equation for shear-thickening materials can also be found elsewhere [16].

$$Q = \pi \frac{3\Delta P_{tot}{}^4 R^4 + 16\tau_0{}^4 L^4 - 8\tau_0 L R^3 \Delta P_{tot}{}^3}{24 \Delta P_{tot}^3 \mu L} \tag{4}$$

As seen from Eq. (4), the yield stress is not the sole dominating factor in the pressure loss, rather it becomes significant when the (τ_0/μ) ratio is high [6].

However, the pressurized pipe flow of concrete is more complicated than expressed in Eq. (4), as the pressure loss predictions based on Eq. (4) overestimate the real pressure losses due to the lack of consideration of the cement paste-rich thin layer formed in the pipe wall region during pumping. This layer, which is rich in water and fine particles, is called the 'lubrication layer', and it is believed that it is formed due to the flow-induced particle migration phenomenon [17, 19] (see further). The properties of the lubrication layer significantly influence the pressure-discharge rate relation.

As long as the shear stress is sufficiently high, the CVC (characterized by relatively low slump values) can also be sheared in the pipelines. The CVC mixtures with relatively low slump values (the yield stress of the mixture is higher than the shear stress at the pipe wall region) can also move inside the pipe as a plug propelled by the lubrication layer at low flow rates. Nevertheless, this requires the formation of the lubrication layer, which in some cases (e.g. ultra-high performance concrete) is not straightforward. The occurrence of plug flow is dependent on the pumping characteristics, pipe diameter and concrete rheological and interfacial properties [20].

Properly designed concrete mixtures must balance these properties to ensure smooth flow through the pipeline without segregation or blockage [3].

2.2 Flow-Induced Particle Migration Phenomenon and Formation of Lubrication Layer

Under the influence of shear forces, particles within the concrete mixture tend to redistribute, with coarser particles moving towards the centre of the pipeline, forcing finer particles to move towards the pipe surface, creating a lubrication layer [21] as schematically illustrated in Fig. 3 [22].

This shear-induced particle migration (SIPM) mechanism is driven by particle size, particle packing, viscosity of the medium and the shear rate applied during pumping. Particle migration during flow can have both positive and negative effects

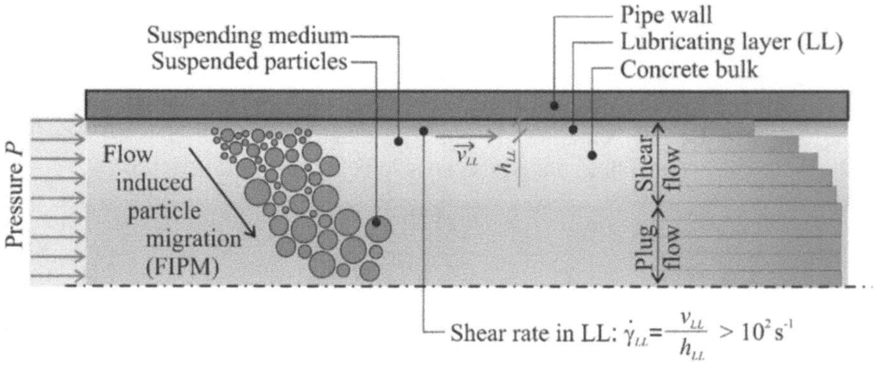

Fig. 3 Schematic illustration of the formation of the lubrication layer due to flow-induced particle migration (the dashed-dotted line is the pipe center) [22]

on the pumpability of concrete. On the one hand, the migration of fine particles and water to the near regions of the pipe wall enhances the flow by reducing the flow resistance and providing a shear plane where the remaining bulk material can relatively move forward by shearing at the boundary [23]. The lubrication layer plays a crucial role in reducing friction and facilitating the flow of concrete [24].

If the applied shear stress at the boundary between the lubrication layer and the concrete is smaller than the yield stress of concrete, the bulk concrete remains unsheared and all shearing occurs in the lubrication layer. However, in case the yield stress of concrete is sufficiently low, the bulk concrete is also partially sheared (as in the case of CVC with relatively low yield stress) or nearly fully sheared (as in the case of self-compacting concrete) as soon as the boundary shear stress exceeds the yield stress of concrete. For a fixed pressure loss, the amount of shearing will not only be dependent on the yield stress of the concrete, but also on its viscosity, and on the properties of the lubrication layer [23]. Therefore, due to the presence of the lubrication layer and partial shear deformation occurring in the bulk concrete, the velocity profile of the concrete differs from the one shown in the purple line in Fig. 2c and becomes the one shown in the yellow line.

The thickness and composition of the lubrication layer depend on the concrete mixture's properties, such as water-cement ratio, aggregate size, and the presence of admixtures. Alekseev [25] and Weber [26] were the first researchers suggesting the existence of this layer. This was followed by many researchers confirming the existence of this layer and its effects on the movement of concrete in pumping pipes [18, 21, 27, 28].

It has been experimentally shown, using particle image velocimetry, that there is no real slip at the surface of the pipe, however, the velocity profile steeply grows in the lubrication layer, showing a thickness of approximately 2 mm (Fig. 4). The material in this layer, with lower yield stress and lower viscosity than the bulk concrete, is intensively sheared. The bulk concrete is partially sheared and/or shows a plug flow, depending on the yield stress of the concrete. It is observed that the effect of the lubrication layer on the overall velocity profile is more pronounced for stiffer concrete [21].

2.3 *Rheology of Pumped Concrete*

The rheological behaviour of concrete during pumping is critical to its performance. As an example, concrete can show shear-thickening or shear-thinning behaviour. This means that its viscosity changes as the shear rate increases. In case of shear-thickening behaviour, which could occur for some types of self-compacting concrete, this could lead to increased pumping pressures.

Furthermore, the rheological properties of the fresh concrete after pumping can be modified due to the pumping process. Some rheological measurements carried out before and after the pumping process showed that pumping increases concrete yield stress and reduces its viscosity [29].

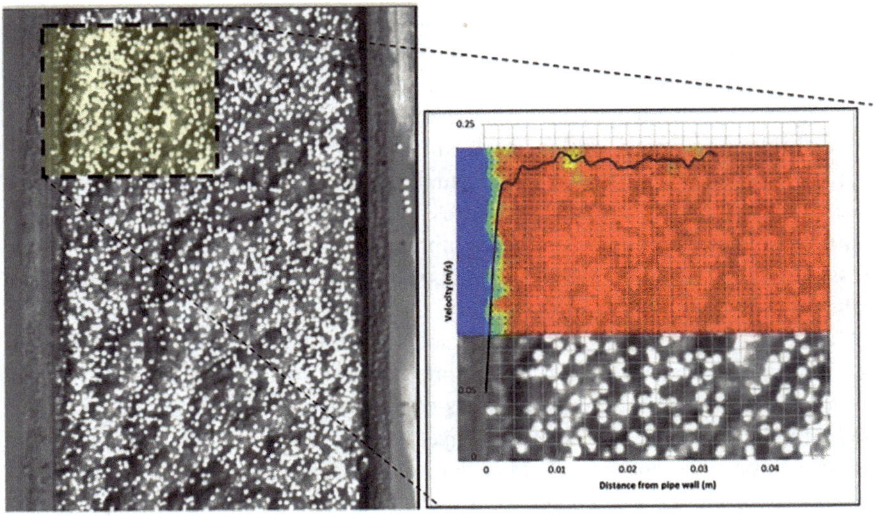

Fig. 4 Lubrication layer observation with particle image velocimetry (PIV) analysis mean (left image captured by the camera, right PIV analysis of the highlighted zone on the left image) [21]

Excessive shear can also lead to particle segregation and bleeding, compromising the quality of the placed concrete. The pumping process may cause significant changes in fresh properties of the concrete, especially air content (see also 3.2).

3 Challenges in Concrete Pumping

3.1 Blockage

Blockages are a common issue in concrete pumping, often caused by improper mix design, inadequate pumping pressure, pipeline irregularities, or a bad startup operation. It has long been known that particle interlocking, as schematically illustrated in Fig. 5 (left), mainly due to the migration of mixing water under pressure or due to aggregate inertia, is the main mechanism of severe blockage, particularly at bends [30]. In early studies, the increase in flow resistance has been associated with the change of concrete from a saturated state to an unsaturated state. Some early researchers proposed a flow resistance versus water-cement ratio relationship for the saturated and unsaturated concretes, Fig. 5 (right) [30]. Unsaturated concrete leads to direct contact between solid particles, turning the material into a friction material, instead of a suspension.

For sure, a decrease in the water-cement ratio will eventually increase the flow resistance of concrete. However, a saturated concrete, as the one suggested to avoid friction, may nevertheless undergo blockage due to the transition of its saturated

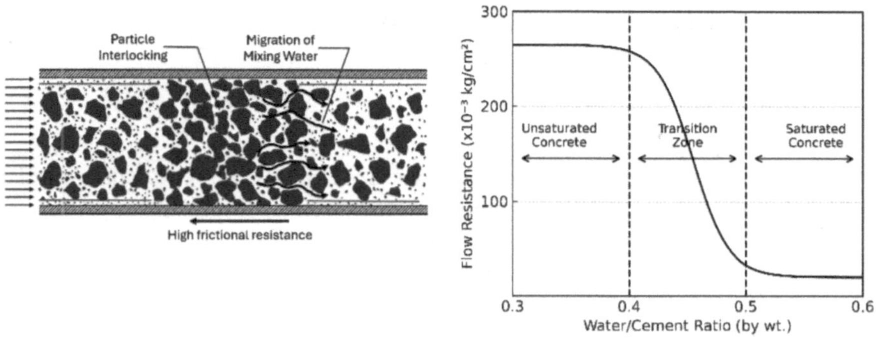

Fig. 5 Dewatering of concrete during pumping causing blockage due to particle interlocking (left) (Redrawn from [30]), flow resistance versus water-cement ratio relationship for different saturation states of concrete (right) (Adopted from [31])

condition into an unsaturated state because of the migration of mixing water under pressure, as schematically illustrated in Fig. 5 (left).

To reduce the friction between the concrete and the inner wall surface of the pipe, a cement-rich paste (grout, also called primer) with a high water-cement ratio is generally pumped before starting the real pumping operation in practice. The primer creates a kind of artificial lubricating layer, reducing the required pumping pressure and the risk of blockage. On the other hand, in some cases, blockage can still occur due to the formation of a dense plug ahead of the flow caused by the coarse aggregate particles leaving the concrete front and starting to flow through the grout section [32]. Concrete mixtures with similar compressive strength showed decreased pumping performance as the maximum coarse aggregate size increased [33]. Reducing the quantity of coarse aggregates and ensuring an adequate grain size distribution of sand are the other aggregate-related measures to prevent blocking.

3.2 *Changes in the Air-Void System of Fresh Concrete*

It has been found that the changes in the air void system induced by pumping are dependent on the particular mix design and discharge conditions at the point of placement [34]. The characteristics of the entrained air and the network of air bubbles may be negatively influenced by the pumping process. When an air-entrained concrete is pumped, the spacing factor is usually increased, and the total amount of air is mostly decreased [35]. However, the situation can be different in some concrete mixtures containing some polyacrylate superplasticizers with air-entraining property. In such mixtures, not the spacing factor, but the total amount of entrained air may increase due to the coalescence of small air bubbles into the coarse air bubbles [35].

4 Recent Advancements in Concrete Pumping Technology

4.1 Smart Pumping Systems

Recent advancements in sensor technologies and data analysis techniques have enabled the development of smart pumping systems. Apart from the pumping characteristics, a few examples can be given for the concepts of the semi-autonomous operation of a mobile concrete pump [36] and control strategies to reduce vibrations induced by the pumping procedure [37].

Some other systems based on monitoring real-time parameters such as pressure, flow rate, and concrete consistency are in the development stage for dynamic adjustments to optimize pumping performance.

4.2 Improved Pipeline Materials

The use of advanced materials, such as ultra-high-molecular-weight polyethylene (UHMWPE), has improved the durability and wear resistance of pumping pipelines. These materials reduce friction and extend the service life of the pipeline.

4.3 Pumping Aids

The incorporation of rheology-modifying admixtures, such as superplasticizers and viscosity-modifying agents usually enhances the pumpability of concrete. Specifically, it is well known that for normal and rich concrete mixtures, water reducers are beneficial in transforming stiff concrete mixtures into plastic and more pumpable mixtures at a given water-cement ratio. This may involve an increase in the sand content of concrete to reduce the risk of bleeding [38]. Furthermore, special admixtures under the name of 'pumping aids' are available in the chemical admixtures market.

5 Active Control Opportunities

As previously mentioned, once the concrete enters the pump, the operator can only passively consider the evolution of the rheological properties of the cementitious material, without any means to actively modify the material properties.

In a first trial to address this limitation, Choi et al. [39] proposed a first pilot test by applying an externally imposed electromagnetic field to improve the formation of the lubrication layer. This technique was inspired by a novel electro-osmosis-based formwork removal technique [40, 41]. Although Choi et al. used carbon steel pipes, which prevented the magnetic field from fully penetrating the bulk of the pumped concrete, the method somehow appeared to improve the lubrication layer formation. The effect has been associated with the activation of the water molecules by the electromagnetic field. The researchers reported an increase in the relative velocity in the lubrication layer after an electromagnetic field application, proposing two possible reasons: (i) excitation of water movement and (ii) reduced friction resistance between the lubrication layer and the wall of the pipe. The reported decrease in total pressure under the same flow rate was around 30%, compared with the original pressure (Fig. 6). It was hypothesized that an improved lubrication layer was formed due to the electromagnetic field provided in the first 20 meters of the pipeline with a total length of 1000 m. The obtained decrease in total pressure was maintained along the pipe during pumping, as the measured pressure losses along the entire pipeline remained almost linear.

In another study, the application of ultrasound has been reported as an effective way to improve pumpability by manipulating the lubrication layer with externally applied ultrasound energy [42].

All studies published until now on the active control of concrete pumping have focused on manipulating the formation of the lubrication layer by an externally imposed electromagnetic field [39] and ultrasound [42].

The recently developed active rheology control (ARC) concept, with successful lab-scale proof of concept [1, 43], seems to be applicable to concrete pumping. The ARC opportunities during pumping have been studied within the ERC Advanced Grant 'SmartCast' project [1] at Ghent University with a closed-loop small-scale

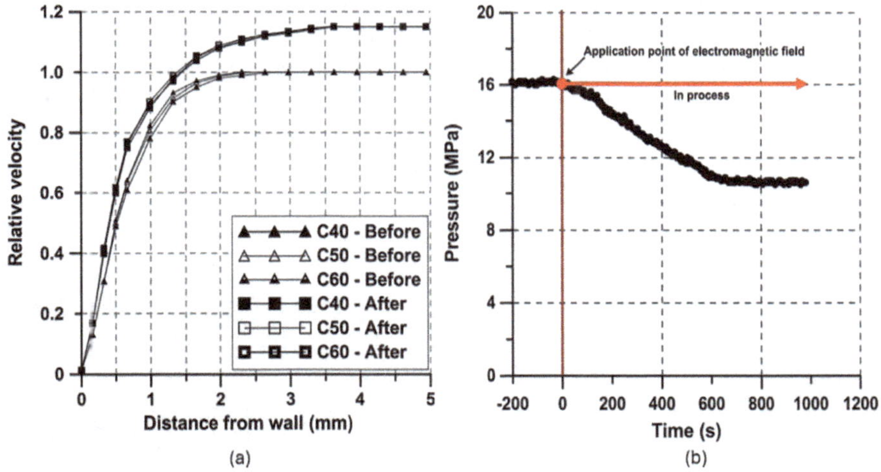

Fig. 6 Normalized velocity profile near the wall of the pipe before and after electromagnetic field (**a**), changes in the measured pumping pressure over time after applying an electromagnetic field for concrete C60 [39] (**b**)

pumping circuit equipped with ultrasound velocity profile (UVP) instrument, pressure and weight sensors. The pipe diameter of this pumping circuit is 25 mm, suitable for pumping tests on mortar mixtures. The UVP instrument in this pumping system enables measuring the velocity profile of cementitious materials flowing in a pipe but typically limited to a depth of the order of 10 mm, which is sufficient to focus on the lubrication layer. The velocity profile measurement of the device is based on the Doppler echo effect through the transducers attached to the outer surface of the pipes [44]. The pipe material in the pumping circuit has been chosen as stainless steel, with low magnetic permeability, so that the externally applied magnetic field is not carried through the pipe itself but is transmitted to the pumped material. This pumping setup allows to evaluate the ARC possibilities in case of pumping, using a magnetic field as the trigger signal.

The possibility of on-demand rheology control while pumping has been shown by De Schryver [45] during his PhD studies within the framework of the ERC Advanced Grant 'SmartCast' project. The active control can be considered as an on-demand triggered thixotropy mechanism, caused by particle clustering governed by the applied magnetic field. To provide a fundamental understanding on the magneto-rheological response of magnetically responsive cementitious mixtures during pumping, an experiment was performed consisting of two pumping situations. In the experiments, a magneto-responsive cementitious mixture, obtained using fly ash particles with magnetic phases, was pumped using the pumping setup shown in Fig. 7a. Two conditions were tested, with and without magnetic control within a 1.5-meter-long pipe section. The magnet control, if present, was provided by means of thin permanent magnets providing a constant magnetic field (~ 0.17 T) perpendicular to the flow direction, Fig. 7b.

The magneto-rheological response was monitored by means of the observed pressure losses in both cases (with and without magnetic control). The effect of the magnetic control was evaluated in such a way that hydration and temperature effects were eliminated, by comparing the behaviour in magnetic and non-magnetic sections within the same test. As shown in Fig. 8, a pure magneto-rheological response has been observed, with higher pressure losses in the case of magnetic control due to magnetic particle cluster formation.

Fig. 7 Pumping system (**a**) and magnetic section (**b**) used for magneto-rheological pumping experiments [45]

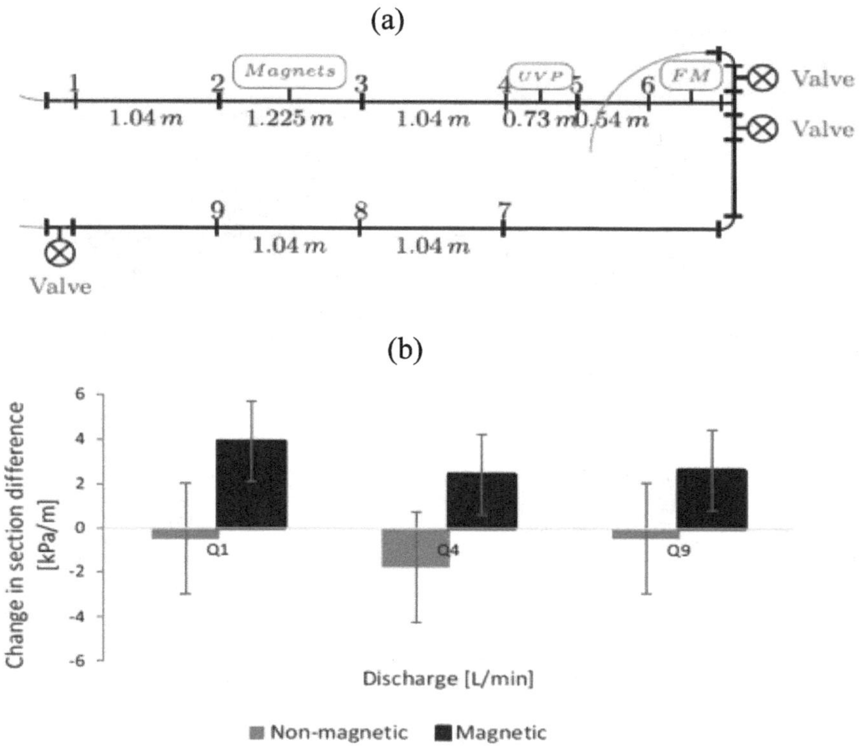

Fig. 8 Plan view of pumping circuit (**a**), changes in pressure loss section difference before and after replacing section 2–3 with a magnetic section (**b**). The pressure loss section difference is calculated as the pressure loss differences between magnetic (2–3, 3–4) and non-magnetic (4–5, 5–6) sections [45]

More specifically, Fig. 8b shows that a significant pressure loss occurs in the section of the pipe where the magnetic field is applied compared to the section where no magnetic field is applied. This expected result originated from the magnetic particles forming columnar structures positioned perpendicular to the magnetic field direction, as evidenced by pictures taken from inside the magnetic portion of the pipeline, Fig. 9.

Although using a sole magnetic section providing a constant static magnetic field perpendicular to flow direction can only provide a kind of stiffening control, the experimentally obtained insights were very instrumental in view of a more general study of Active Rheology Control (ARC). In [45], the magnetic segment of the pumping circuit was composed of permanent magnets providing a fixed magnetic field oriented perpendicular to the flow direction, with no means of adjusting the properties of the magnetic field during flow. A further valorisation project to build a prototype device for active control of fresh concrete rheology is still ongoing at Ghent University [46], involving electromagnetic coils that can deliver alternating

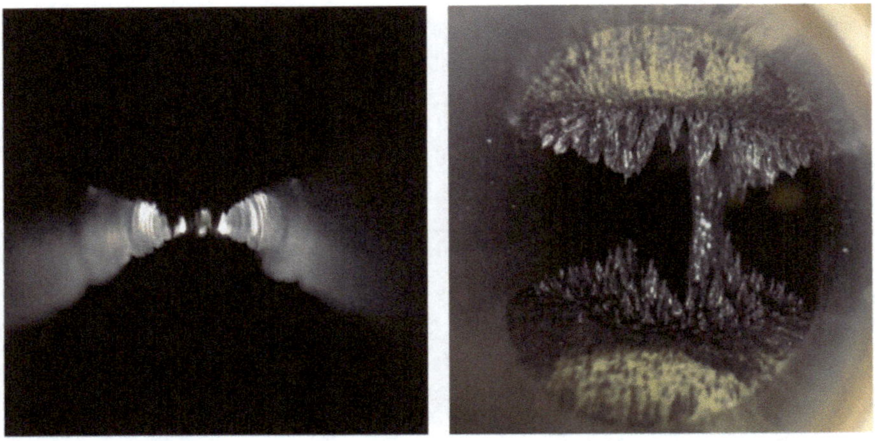

Fig. 9 Magnetic columnar structures inside the magnetic portion of the pipeline [45]

magnetic fields. The studies in [46] fundamentally aim to improve lubricating layer properties by applying a pulsating magnetic field to vibrate the magnetic particles contained within the cementitious material in the lubrication layer zone. While the results of the studies in [46] cannot be shared here yet due to project confidentiality constraints, it was observed that the pulsating magnetic field had a significant impact on the pumping pressure. Findings from this project also suggest that, in addition to the type of magnetic field, the characteristics of the electromagnetic signal influence the pump pressure. First results showed that when a unipolar signal was used at a certain frequency, a strong micro-vibration was obtained. However, the pressure loss difference between the magnetized and unmagnetized pumping pipe sections was also increased, due to a magnetic clustering effect dominating the positive influence of the magneto-induced particle vibration. The challenge is to make the particle vibration dominant instead of the particle clustering.

6 Summary and Outlook

Pumping fresh concrete is a very popular technique, daily used in concrete industry. Although the principle seems simple, the behaviour of the fresh concrete within the pumping pipes can be complex, and results in several challenges. For a constant pressure loss, the shear stress distribution within the pumping pipes always follows a linear evolution from pipe axis to pipe surface, independent from the rheological properties of the fresh concrete. However, the shear rate distribution significantly depends on the material, and on phenomena like shear-induced particle migration. In principle, by modifying the rheological properties of fresh concrete, the pressure-discharge relation can be modified as well. Active rheology control techniques, as recently introduced, can be considered to facilitate pumping operations, e.g. based on magneto-rheological principles. First literature results have been summarized. Further research is ongoing to acquire better insight into the competition between micro-vibration and cluster formation, and to bring the pumping control technology closer to practical applications.

References

1. De Schutter, G., Lesage, K.: Active Rheology Control of Cementitious Materials, Modern Concrete Technology 23. CRC Press, Taylor and Francis Group (2023)
2. Kim, K.Y., Kim, Y.J., Choi, M.S.: Horizontal-vertical ratio for concrete pumping pipe. Case. Stud. Constr. Mat. **16** (2022)
3. Mikhalev, D., Nejad, S.F., Ng, S., et. al., Practical Insights and Advances in Concrete Pumping, RILEM Technical Letters (2024) 9: 1–9 (2024)
4. ACI 304.2R-96: Placing Concrete by Pumping Methods. American Concrete Institute (2008)
5. Roussel, N., Coussot, P.: "fifty-cent rheometer" for yield stress measurements: from slump to spreading flow. J. Rheol. **49**(3), 705–718 (2005)
6. De Schutter, G., Feys, D.: Pumping of fresh concrete: insights and challenges. RILEM Technical Letters. **1**, 76–80 (2016)
7. Feys, D.: Interactions between Rheological Properties and Pumping of Self-Compacting Concrete., PhD thesis,. Ghent University (2009)
8. Secrieu, E.: Pumping Behaviour of Modern Concretes—Characterisation and Prediction, PhD Thesis,. Technischen Universität Dresden (2018)
9. Wallevik, O.H., Wallevik, J.E.: Rheology as a tool in concrete science: the use of rheographs and workability boxes. Cement Concrete Res. **41**(12), 1279–1288 (2011)
10. Feys, D., De Schutter, G., Verhoeven, R.: Parameters influencing pressure during pumping of self-compacting concrete. Mater. Struct. **46**(4), 533–555 (2013)
11. Feys, D., De Schutter, G., Fataei, S., Martys, N.S., Mechtcherine, V.: Pumping of concrete: understanding a common placement method with lots of challenges. Cement. Concrete. Res. **154** (2022)
12. Macosko, C.W.: Rheology: Principles, Measurements, and Applications. Wiley-VCH, Inc. (1994)
13. Mezger, T.G.: The Rheology Handbook for Users of Rotational and Oscillatory Rheometers, 4th edn. Vincentz Network, Hanover, Germany (2014)
14. Malkin, A.Y., Isayev, A.I.: Rheology - Concept, Methods, and Applications, 3rd edn. ChemTec Publishing (2017)
15. Buckingham, E.: On plastic flow through capillary tubes. Proc. Am. Soc. Testing Materials, 1154–1156 (1921)
16. Feys, D., Verhoeven, R., De Schutter, G.: Extension of the Poiseuille formula for shear-thickening materials and application to self-compacting concrete. Appl. Rheol. **18**(6) (2008)
17. Jacobsen, S., Haugan, L., Hammer, T.A., Kalogiannidis, E.: Flow conditions of fresh mortar and concrete in different pipes. Cement Concrete Res. **39**(11), 997–1006 (2009)
18. Fataei, S., Secrieu, E., Mechtcherine, V.: Experimental insights into concrete flow-regimes subject to shear-induced particle migration (SIPM) during pumping. Materials. **13**(5) (2020)
19. Salinas, A., Feys, D.: Estimation of lubrication layer thickness and composition through reverse engineering of Interface Rheometry tests. Materials. **13**(8) (2020)
20. Feys, D., Khayat, K.H., Perez-Schell, A., Khatib, R.: Prediction of pumping pressure by means of new tribometer for highly-workable concrete. Cement Concrete Comp. **57**, 102–115 (2015)
21. Le, H.D., Kadri, E.H., Aggoun, S., Vierendeels, J., Troch, P., De Schutter, G.: Effect of lubrication layer on velocity profile of concrete in a pumping pipe. Mater. Struct. **48**(12), 3991–4003 (2015)

22. Secrieru, E., Khodor, J., Schrofl, C., Mechtcherine, V.: Formation of lubricating layer and flow type during pumping of cement-based materials. Constr. Build. Mater. **178**, 507–517 (2018)
23. Feys, D.: How much is bulk concrete sheared during pumping? Constr. Build. Mater. **223**, 341–351 (2019)
24. Kaplan, D., de Larrard, F., Sedran, T.: Design of concrete pumping circuit. ACI Mater. J. **102**(2), 110–117 (2005)
25. Aleekseev, S.N.: On the calculation of resistance in the pipes of concrete pumps. Mekhanizat-siyaStroitel'stva. **9**(1), 8–13 (1952)
26. Weber, R.: The Transport of Concrete by Pipeline. Cement and Concrete Association, London (1968)
27. Choi, M.S., Kim, Y.J., Kwon, S.H.: Prediction on pipe flow of pumped concrete based on shear-induced particle migration. Cement Concrete Res. **52**, 216–224 (2013)
28. Xie, X.Y., Zhang, L.H., Shi, C.J., Liu, X.M.: Prediction of lubrication layer properties of pumped concrete based on flow induced particle migration. Constr. Build. Mater. **322** (2022)
29. Secrieru, E., Cotardo, D., Mechtcherine, V., Lohaus, L., Schröfl, C., Begemann, C.: Changes in concrete properties during pumping and formation of lubricating material under pressure. Cement Concrete Res. **108**, 129–139 (2018)
30. Browne, R.D., Bamforth, P.B.: Tests to establish concrete pumpability. ACI J. **74**, 193–203 (1977)
31. Ede, A.N.: The resistance of concrete pumped through pipelines. Mag. Concr. Res. **9**(27), 129–140 (1957)
32. Kaplan, D., de Larrard, F., Sedran, T.: Avoidance of blockages in concrete pumping process. ACI Mater. J. **102**(3), 183–191 (2005)
33. Kwon, S.H., Jang, K.P., Kim, J.H., Shah, S.P.: State of the art on prediction of concrete pumping. Int J Concr Struct M. **10**(3), S75–S85 (2016)
34. Vosahlik, J., Riding, K.A., Feys, D., Lindquist, W., Keller, L., Van Zetten, S., Schulz, B.: Concrete pumping and its effect on the air void system. Mater. Struct. **51**(4) (2018)
35. P.C. Aitcin, R.J. Flatt, Science and Technology of Concrete Admixtures, Woodhead Publishing2016
36. Meiringer, M., Kugi, A., Kemmetmüller, W.: Semi-autonomous operation of a mobile concrete pump. Automat Constr. **156** (2023)
37. Ripamonti, F., Bucca, G., Fava, V., Resta, F.: A smart-hose for concrete displacing booms. Proc. SPIE. **9803** (2016)
38. Ramachandran, V.S.: Concrete Admixtures Handbook: Properties, Science and Handbook. Noyes Publications (1995)
39. Choi, M.S., Kim, Y.S., Kim, J.H., Kim, J.S., Kwon, S.H.: Effects of an externally imposed electromagnetic field on the formation of a lubrication layer in concrete pumping. Constr. Build. Mater. **61**, 18–23 (2014)
40. Goudjil, N., Djelal, C., Vanhove, Y., Kada, H., Heloun, N.: Impact of temperature on the demoulding of concrete elements with a polarization process. Constr. Build. Mater. **54**, 402–412 (2014)
41. Goudjil, N., Vanhove, Y., Djelal, C., Kada, H.: Electro-osmosis applied for formwork removal of concrete. J. Adv. Concr. Technol. **10**(9), 301–312 (2012)
42. Choi, M.S., Kim, Y.S., Kim, Y.U.: Effect of ultrasound on the formation of a lubrication layer in concrete pumping. J. Adv. Concr. Technol. **14**(3), 95–101 (2016)
43. De Schutter, G., Lesage, K.: Active control of properties of concrete: a review. Mater. Struct. **51**(5) (2018)
44. Wiklund, J., Shahram, I., Stading, M.: Methodology for in-line rheology by ultrasound Doppler velocity profiling and pressure difference techniques. Chem. Eng. Sci. **62**(16), 4277–4293 (2007)
45. De Schryver, R.: Active Rheology Control of Cementitious Materials: Numerical and Experimental Pumping Investigation., PhD Thesis,. Ghent University (2022)
46. De Schutter, G., Matthys, S., Yardimci, M.Y.: Industial-scale prototyping, validation and demonstration of a magnetic device for active rheology control of concrete during pumping, industrial research fund (IOF) advanced project. Ghent University.

Active Control of Formwork Casting of Cementitious Materials

Angela Tetteh Tawiah and Geert De Schutter

Abstract Formwork plays a vital role in concrete construction, serving as a temporary or permanent mould that holds fresh concrete in place until it hardens and reaches the desired strength. It provides essential support, maintains the shape of the structure, and ensures the concrete cures properly, all of which are crucial for the quality, safety, and efficiency of a construction project. This chapter briefly recapitulates the types of formworks and their characteristics. Furthermore, the chapter lists some potential problems occurring with formwork and introduces how active control during formwork casting can be useful to overcome some of the challenges related to formwork casting.

Keywords Concrete · Formwork · Active control

1 Introduction

Formwork plays a vital role in concrete construction, serving as a temporary or permanent mould that holds fresh concrete in place until it hardens and reaches the desired strength [1]. It provides essential support, maintains the shape of the structure, and ensures the concrete cures properly, all of which are crucial for the quality, safety, and efficiency of a construction project. This process is vital for achieving the precise geometries required in modern architectural designs. The selection of appropriate formwork systems is influenced by factors such as structural geometry, construction speed, cost, and surface finish quality [2].

Well-designed formwork ensures dimensional accuracy and helps achieve complex architectural designs, from simple slabs to intricate curved elements. It

A. T. Tawiah
University of Padova, Padova, Italy

G. De Schutter (✉)
Magnel-Vandepitte Laboratory, Department of Structural Engineering and Building Materials, Ghent University, Ghent, Belgium
e-mail: Geert.DeSchutter@ugent.be

G. De Schutter and J. Sanjayan (eds.), *Active Control of Properties of Fresh and Hardening Cementitious Materials*, RILEM State-of-the-Art Reports 50,
https://doi.org/10.1007/978-3-032-30340-0_7

withstands the pressure and weight of fresh concrete, preventing deformation or collapse, while also providing a safe working platform for labourers during pouring and finishing operations. In addition to supporting structural integrity, formwork influences surface quality, for instance, smooth and precise moulds reduce the need for additional finishing work. Modern formwork systems, like modular and reusable panels made from materials like steel, aluminium, or plastic, speed up construction, lower labour costs, and promote sustainability by reducing material waste. Ultimately, formwork is not just a temporary structure but a critical component that impacts the safety, speed, and quality of concrete construction. Careful planning, proper material selection, and timely removal are essential to maximizing its effectiveness and ensuring the project's long-term success. This chapter gives a basic understanding on the active control properties of fresh concrete during formwork casting.

2 Fundamentals of Formwork Casting

2.1 Types of Formworks and Their Characteristics

Steel Formwork

Steel formwork is widely utilized in large-scale construction projects and in structures where a high-quality surface finish is required. Its principal advantages lie in its exceptional durability and long service life, which enable repeated use over numerous projects and contribute to overall cost-effectiveness despite the relatively high initial investment. Steel formwork produces smooth and uniform concrete surfaces, thereby reducing the extent of post-construction finishing operations. However, its significantly greater weight compared to timber or plywood necessitates the use of mechanical lifting equipment during installation, which may increase labour and equipment costs. Nevertheless, its strength, reusability, and ability to deliver superior finishes make steel formwork a highly suitable option for demanding construction applications.

Aluminium Formwork

Aluminium formwork, like steel but considerably lighter, is employed in projects where speed and efficiency are of primary importance. Its resistance to corrosion enhances durability and extends its service life, while its relatively light weight allows for faster erection and dismantling, thereby accelerating construction schedules. Although aluminium formwork involves a high initial investment, this cost is offset by its reusability and the reduction in labour expenses associated with its handling and installation. However, its rigidity makes it less adaptable to complex shapes compared to timber-based alternatives, which may limit its versatility in certain construction contexts.

Timber Formwork

Timber formwork is one of the oldest and most widely utilized types of formworks, particularly suited to small-scale projects or structures requiring complex customization. Its inherent adaptability allows it to be shaped into a wide variety of forms and dimensions, making it highly effective for projects that demand flexible construction solutions. A notable advantage of timber formwork is that it can be fabricated directly on-site, permitting adjustments to be made during the construction process [3]. The availability of timber as a material also contributes to its cost-effectiveness in smaller-scale applications [4]. However, the assembly and dismantling processes are labour-intensive, and the material itself is prone to warping, swelling, and shrinkage due to moisture and environmental conditions [5], which may compromise durability and reduce its service life.

Plywood Formwork

Plywood formwork is constructed from resin-bonded sheets fixed to timber frames, providing a smooth and uniform surface that minimizes the need for subsequent finishing operations. Owing to its relatively low weight, plywood formwork is easier to transport, handle, and install, thereby enhancing efficiency on construction sites. When properly handled and maintained, it is capable of multiple reuse cycles, contributing to cost-effectiveness in building projects. Despite its inherent susceptibility to moisture, swelling, and biological or chemical degradation, appropriate treatment significantly improves its resistance to these factors, thereby extending its service life [6]. In addition to its economic advantages, plywood formwork offers favourable performance characteristics such as ease of workability, thermal insulation, and reduced safety risks associated with manual handling. These attributes underscore its suitability as a practical and efficient formwork material within diverse construction environments.

Plastic and Polymer Formwork

Plastic formwork systems consist of modular interlocking panels manufactured from lightweight yet durable plastic materials. They are designed for repeated use with minimal wear, thereby ensuring cost efficiency over multiple construction cycles. In addition, plastic formwork produces consistent and smooth concrete surface finishes, reducing the need for extensive post-construction treatment. Despite these advantages, its prefabricated nature limits flexibility, as the panels may not be suitable for projects requiring complex or irregular geometries.

Fabric Formwork

Fabric formwork represents an innovative approach in concrete construction, employing high-strength fabrics to mould concrete into complex and non-linear geometries. This method enables the production of structurally efficient forms by reducing material use through optimized curved and irregular shapes. However, such designs necessitate advanced structural analysis methods, including three-dimensional modelling, rather than traditional sectional analysis typically applied to prismatic members [7]. From an architectural perspective, fabric formwork enhances design freedom by facilitating fluid and organic forms that are not achievable with rigid systems. Nevertheless, its successful application requires careful planning and specialized expertise, which may increase both design and labour costs. Furthermore, reusability of the fabric depends on the specific material and project requirements, potentially limiting its practicality in certain contexts.

2.2 Potential Problems Occurring with Formworks

Insufficient Flowability or Filling Capacity

Non flowable concrete may not fill the formwork completely and this can create structural defects, weaken the concrete, and lead to costly repairs or safety risks. This insufficient flowability can be due to the low water-cement ratio or absence of appropriate chemical admixtures that can help increase fluidity in the fresh concrete [8]. In addition, inadequate compaction during placement prevents concrete from flowing into every crevice in the formwork especially around reinforcement and corners. This problem may lead to honeycombing, voids, and cold joints because of incomplete filling between layers of concrete causing weak connections and surface defects [9].

Too High Formwork Pressure

Fresh concrete exerts significant lateral pressure on the formwork, which can lead to formwork deformation or collapse if the pressure becomes too high. The magnitude of this pressure is influenced by several factors, including the rate of placement, concrete temperature, mix consistency, and the height of the pour [10]. Highly flowable mixes such as self-compacting concrete (SCC) exert greater hydrostatic pressure due to their enhanced fluidity and slower setting characteristics [11]. Too high formwork pressure can cause formwork panes to bend, burst or collapse. Pressure-induced movement can also lead to structure deformations. To control the formwork pressure, pour height can be monitored as well as pressure sensors can be used for real time adjustments during casting [12].

Leaking of Formwork

Formwork leakage occurs when fresh concrete escapes through gaps or weak points in the formwork system. Three key factors influence this issue: gaps between formwork joints, material properties of formwork and the pressure exerted by the plastic state of the concrete. Leakage pathways, such as cracks, panel separations, or unsealed tie holes, allow cementitious material to flow out. These gaps can result from poor fabrication, construction damage, or inadequate tightening of formwork panels. The width of the gap between formwork panels influences both the flow characteristics of the escaping material and the amount and composition of the leaked concrete [13].

While sealing methods like rubber gaskets, foam strips, and caulking compounds help minimize leakage, achieving perfect connections remains a challenge.

The flexibility and stability of formwork materials affect their ability to withstand leakage. If the formwork is too flexible or improperly installed, displacement may occur, increasing the likelihood of leaks. The self-weight of fresh concrete generates vertical forces that transfer pressure to the formwork walls. Under high pressure, mortar can seep through small openings at the concrete–formwork interface. Pressure begins to decrease as the concrete develops an internal microstructure, initially due to physical factors such as internal friction, and later through chemical reactions during setting.

Demoulding Issues

Demoulding refers to the process of removing the formwork after the concrete has set sufficiently. Several issues can arise during demoulding, affecting the quality of the concrete surface and structural integrity. Surface damages, cracks or deformation of concrete can happen due to insufficient strength development at the time of formwork removal. The appropriate time for demoulding depends on factors such as cement type, curing conditions and ambient temperature [14]. Strong adhesion between concrete and formwork can also lead to chipped edges and surface defects. This is especially common when using absorbent or rough- surfaced formwork.

3 Active Control of Fresh Concrete Properties During Formwork Casting

3.1 Workability and Flowability Adjustments (Rheology Control)

Formwork filling can be very challenging, especially in case of tall and slender concrete walls or geometrically complex elements with dense reinforcement. Low quality placement, with improper filling of the formwork can lead to important mechanical or durability problems. Incomplete encapsulation of reinforcing bars leads to reduced bond strength, and potentially insufficient cooperation between the reinforcement bars and the concrete. Entrapped air will reduce the overall strength and load bearing capacity of the element, as well as the overall durability of the structure because of faster transport of potentially aggressive fluids and gases into the concrete. Especially a perfect placement of the cover zone is crucial to the overall performance of the element, as any quality loss due to improper placement is directly hampering the long-term protection of the steel reinforcement.

In a passive way, a first and indispensable aspect of the proper placement of concrete during formwork casting is a fit-for-purpose mix design. The correct consistency should be aimed at, in view of the specific element to be cast with the considered placement technique. In case of traditional concrete, this typically consists of an adequate dosage of superplasticizers to reach target slump values. One step further, self-compacting concrete could be designed, combining latest generation superplasticizers in combination with an adapted particle structure design including less coarse particles and more fines. Self-compacting concrete originally has been first developed in Japan in view of challenges with formwork casting of concrete and resulting durability issues [15].

In an active way, different vibration techniques have been applied already for decades in concrete industry. The most used method is the application of handheld poker vibration, consisting of an excentric mass rotating at high frequency. As already mentioned in Chap. 1, vibration has a certain range of action, depending on frequency and amplitude, triggering the (solid) particles in the cementitious material to respond (vibrate). As a result, the yield stress will be strongly reduced within the affected zone, liquefying the material in a sufficient way to fill the formwork [16]. As an alternative to poker vibration, formwork vibration can also be applied. This is more common in precast industry, e.g. producing prefabricated beams or concrete elements such as pipes, lintels and so on. A well-designed positioning of the formwork vibrators is needed, to reach proper vibration conditions in all zones of the cast element. Due to interference of mechanical waves, the action of different vibrators could locally cancel out each other, leaving some zones with improper compaction.

Because of issues with health and environment, mechanical vibration is not always the best and certainly not the most liked option to actively improve the flowability while casting concrete. The absence of vibration and noise is another major advantage of the use of self-compacting concrete. However, as vibration can then no longer be used as a method to alter the workability while casting, as it would even lead to segregation of the self-compacting concrete, other ways are needed to intervene. For this reason, the concept of active rheology control has recently been developed [13], as already introduced in Chap. 3. One option is to rely on switchable superplasticizers that can be triggered by an external signal to modify the rheological properties of the fresh concrete while processing. While this option has been successfully implemented in a laboratory scale proof of concept, further upscaling steps are needed to apply it to formwork casting.

3.2 Form Pressure and Stability (Stiffening Control)

Once the concrete is in final position within the formwork, a fast reduction in workability is helpful for a fast reduction of the formwork pressures. In a natural way, the loss of workability is linked to the internal structural formation and clustering of particles at micrometre level, at least partly due to the formation of initial hydration products [17]. Cementitious materials with increased thixotropy behaviour can lead to a faster reduction of formwork pressure [17]. In a passive way, this could be improved by an optimized mix design, e.g. including nano clay particles into the mixture [18].

In an active way, the internal structural formation or clustering of particles could be triggered e.g. in a magnetic way by applying an external magnetic field to a cementitious material containing magnetizable particles [19], as already introduced in Chap. 4 in view of on-demand setting. The applicability of this control feature has been shown already at laboratory scale [13]. Further upscaling steps have recently been studied in view of 3D Concrete Printing [20], as also discussed in Chap. 8. Further application of this active setting control principle to the case of formwork casting has nevertheless not yet been reported up to now.

3.3 Form Leakage Control

As formwork is typically composed of a combination of separate panels or elements (see also sect. 7.2), the presence of formwork joints is often unavoidable. Especially in case of very flowable concrete, including self-compacting concrete, the risk of material flowing out of the formwork through the joints can be high. This can lead to local or even more extended defects, with concrete zones showing loss of paste resulting in porous clusters of aggregate particles.

In a passive way, a good tightness control of the formwork is desirable, as well as the design of a stable mix design. Higher viscosity and/or higher thixotropy might be useful to reduce the risk of formwork leakage, although it is not the preferred choice to facilitate formwork casting. Because of the contradictory requirements, active control options to reduce the risk of formwork leaking can be helpful.

One recently studied option to actively control formwork leakage is based on the magnetic control of cementitious materials containing magnetizable particles [13]. The principle is based on on-demand setting approaches, as already introduced in chap. 4. In the case of formwork leakage control, the magneto-controlled setting is locally applied in the region of the formwork joint, by locally positioned permanent magnets. Some metallic formwork systems have magnet strips near the edges, to facilitate mounting of the formwork panels because of the magnetic forces between the panels. In such a case, the presence of the magnetic strips can also be successfully applied in view of the desired formwork leakage control.

As explained in more detail in [21], based on the principle of active stiffening control (ASC) of cementitious materials, a magnetic field-based active stiffening control methodology has recently been successfully introduced to reduce formwork leakage under pressure. During the experimental studies, a newly developed pressure-driven flow test has been considered. "*A pressurized cylinder with a flow outlet at the bottom was used to mimic formwork leakage under pressure* (see Fig. 1). *The flow outlet has a rectangular cross-section, with dimensions 2 mm × 20 mm, and height of approximately 7 mm. Five overhead pressure levels of approximately 10, 20, 30, 40, and 50 kPa were used in this study. For the magnetic field application, the bottom plate of the test cylinder was replaced with a plate containing static magnets with a similar outflow geometry to the non-magnetic case. The permanent magnets create a magnetic field perpendicular to the flow direction.*"

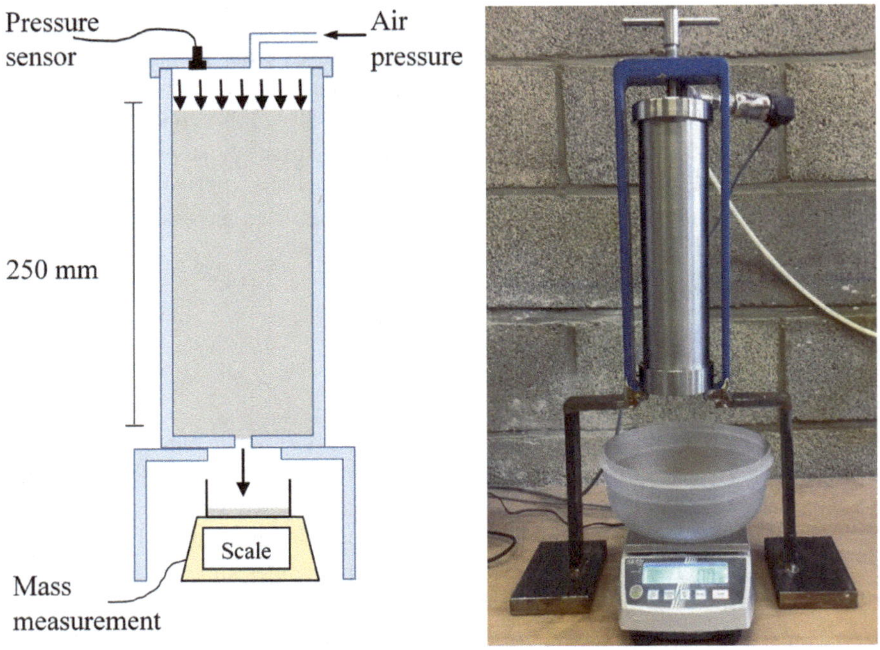

Fig. 1 Test-setup to study pressure-driven flow of cementitious paste [21]

By means of this newly developed pressure-driven flow test, a continuous time recording of the mass of the outflowing material was applied to cement paste containing magnetizable Fe_3O_4 nanoparticles. The accumulated mass of the outflown material can be related to the initially present mass and expressed by the percentage mass loss (M_{loss}). Some representative curves showing this mass loss percentage as a function of time are shown in Fig. 2.

When no magnetic field is applied, all material is flowing out of the setup, through the joint opening in the bottom. In the figures, this is visible as the linear mass loss as a function of time, until reaching a plateau after a certain time, corresponding to complete outflow of all material. Only a small amount of material remains within the test column and is left behind in dead corner zones at the bottom. Alternatively, in presence of a magnetic field with strength of 0.65 T, the mass loss rate is substantially reduced (the linear parts of the curves are less steep), and a final plateau is gradually reached at significantly lower values of the mass loss. The overall reduced mass loss at the plateau level is due to the blockage of the outflow, induced by the magnetic field forcing the magnetizable particles within the cementitious material to internally form some magnetic clusters that will gradually block the outflow opening (representing the formwork joint). The cited reference further links the observed blockage behaviour with a more detailed study of the magnetorheological properties of the cementitious material containing magnetizable nanoparticles.

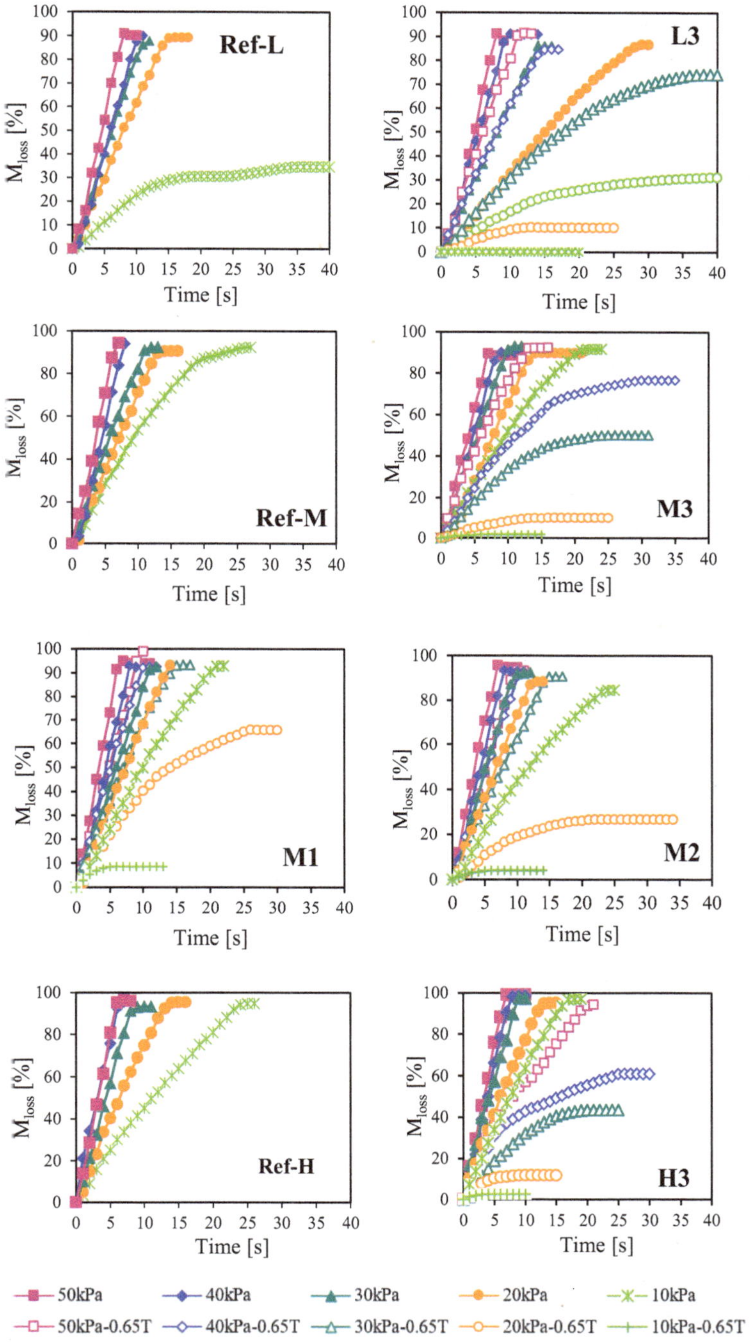

Fig. 2 Mass loss percentage due to overhead pressure as a function of time, with and without magnetic field. Pressure levels are indicated in the figure legend. In case of magnetic field, the name includes 0.65 T, referring to the corresponding magnetic field strength field [21]

As summarized in the cited reference [21], "*The effects of Fe_3O_4 nanoparticle concentration, paste rheology, and applied pressure on the magnetic field-based formwork leakage control are investigated. The experimental results showed that the Active Stiffening Control (ASC) method is effective in reducing the flow rate for formwork pressures of 10-50 kPa and completely blocking the flow at 10-20 kPa. The ASC method was found to be more beneficial for the more flowable pastes that generally exhibit more leakage in practice. The efficacy of the ASC method is also significantly influenced by the Fe_3O_4 nanoparticles concentration. The required amount of Fe_3O_4 nanoparticles for flow reduction should thus be selected accordingly. It is concluded that magneto-responsive additives are a suitable means in applying active stiffening control for formwork leakage reduction.*"

An upscaling of this concept of active formwork leakage control to a real construction environment, either on site or in precast industry, remains to be done.

3.4 Form Demoulding Control

Another point of attention in case of formwork casting is the demoulding control. In current practice, proper preparation of the formwork consists of the application of a demoulding oil before casting the concrete. This is helpful for a proper interaction between the cement paste of the concrete and the formwork, leading to a good quality of the concrete surface with a limited number and size of defects (air bubbles, sand stripes). Furthermore, the demoulding oil reduces the bond between cement paste and formwork, and thus facilitates demoulding, reducing the risk of damaging the surface.

In an active way, the effect of demoulding agent could be replaced by the application of an electric field of limited voltage (e.g. 5 V) polarizing the zone of the concrete in contact with the formwork surface. This technology has been introduced by Goudjil et al., as described in more detail in [22, 23]. Since their studies, now about 15 years ago, no further research papers have been published with more recent results. Application of this technology on real construction sites has also not been reported yet.

4 Summary

Formwork-based casting is the most widely used method for producing concrete structures and elements in modern construction. The selection of formwork materials ranging from timber and plywood to steel, aluminium, plastic, and fabric has a direct impact on cost, efficiency, durability, and the quality of the finished structure. Despite their advantages, all formwork systems face common challenges, including issues related to filling of the formwork, formwork pressure, formwork leakage, and formwork demoulding. To address these challenges, recent advancements have introduced active control technologies such as active rheology control to regulate concrete flow, active stiffening to improve form stability, magneto-rheological leakage control to prevent material loss, and electric demoulding control to enable safer and more efficient form removal. These innovations represent a significant step toward enhancing the performance and reliability of formwork systems in contemporary construction practice.

References

1. Nawi, M., et al.: Suitability of IBS formwork system in Malaysian construction industry. Aust. J. Basic Appl. Sci. **7**(7) (2015)
2. Peurifoy, R.L., Oberlender, G.D.: Formwork for Concrete Structures. McGraw-Hill (2011)
3. Smith, G.R., Hanna, A.S.: Factors influencing formwork productivity. Can. J. Civ. Eng. **20**(1), 144–153 (1993). https://doi.org/10.1139/l93-015
4. Pronk, A., Brancart, S., Sanders, F.: Reusing timber formwork in building construction: testing, redesign, and socio-economic reflection. Urban Plan. **7**(2), 81–96 (2022). https://doi.org/10.17645/up.v7i2.5117
5. Pastori, S., Sergio Mazzucchelli, E., Wallhagen, M.: Hybrid timber-based structures: a state-of-the-art review. Constr. Build. Mater. **359**, 129505 (2022). https://doi.org/10.1016/j.conbuildmat.2022.129505
6. Shah, M.S.A., Noor, N.M., Kueh, A.B.H., Tamin, M.N.: A review on wooden formwork for concrete casting. IOP Conf Ser Mater Sci Eng. **513**, 012036 (2019). https://doi.org/10.1088/1757-899X/513/1/012036
7. Hawkins, W.J., et al.: Flexible formwork technologies—a state-of-the-art review. Struct. Concr. **17**(6), 911–935 (2016). https://doi.org/10.1002/suco.201600117
8. Neville, A.: Properties of Concrete, Fifth edn. Prentice Hall (2012)
9. Mehta, P.K., Monteiro, P.J.M.: Concrete: Microstructure, Properties, and Materials. McGraw-Hill (2006)
10. Li, W., Lin, X., Bao, D.W., Min Xie, Y.: A review of formwork systems for modern concrete construction. Structure. **38**, 52–63 (2022). https://doi.org/10.1016/j.istruc.2022.01.089
11. Khayat, K.H.: Workability-testing-and-performance-of-self-consolidating-concrete. ACI Mater. J. **96**, 346–353 (1999)
12. Craipeau, T., Toussaint, F., Perrot, A., Lecompte, T.: Experimental approach on a moving formwork. Constr. Build. Mater. **270**, 121472 (2021). https://doi.org/10.1016/j.conbuildmat.2020.121472
13. De Schutter, G., Lesage, K.: Active Rheology Control of Cementitious Materials. Crc press (2023)
14. Du, W., Qian, C., Xu, Y., Wei, M., Peng, B., Xie, Y.: Effects of ambient temperature, formwork type, and demoulding time on the thermal deformation of sidewall concrete in underground engineering: experiment, simulation, and engineering practice. Constr. Build. Mater. **353**, 129083 (2022). https://doi.org/10.1016/j.conbuildmat.2022.129083
15. De Schutter, G., et al.: Self-Compacting Concrete. CRC Press, Whittles Pub (2008)
16. G. Grampeix, N. Roussel, J. Dupoirier.: Internal vibration and viscous concrete: application and prediction of the radius of action
17. Roussel, N., Ovarlez, G., Garrault, S., Brumaud, C.: The origins of thixotropy of fresh cement pastes. Cem. Concr. Res. **42**(1) (2012). https://doi.org/10.1016/j.cemconres.2011.09.004
18. Qian, Y., De Schutter, G.: Enhancing thixotropy of fresh cement pastes with nano clay in presence of polycarboxylate ether superplasticizer (PCE). Cem. Concr. Res. **111** (2018). https://doi.org/10.1016/j.cemconres.2018.06.013

19. Zhang, Y., et al.: Active control of thixotropy of magneto-responsive cementitious materials with the intervention of time-varying magnetic fields. Cem. Concr. Res. **181**, 107520 (2024). https://doi.org/10.1016/j.cemconres.2024.107520
20. Zhang, Y., Tao, Y., Zhang, Y., Van Tittelboom, K., Lesage, K., De Schutter, G.: Upscaling active rheology control to cement mortar with the intervention of an inline magnetic field. Constr. Build. Mater. **425**, 135853 (2024). https://doi.org/10.1016/j.conbuildmat.2024.135853
21. Chibulu, C., Jiao, D., Yardimci, M., De Schutter, G.: Magneto-rheology control of cement paste containing Fe3O4 nanoparticles in view of reducing or preventing formwork leakage. Cem. Concr. Compos. **142**, 105176 (2023). https://doi.org/10.1016/j.cemconcomp.2023.105176
22. Goudjil, N., Vanhove, Y., Djelal, C., Kada, H.: Electro-osmosis applied for formwork removal of concrete. J. Adv. Concr. Technol. **10**(9), 301–312 (2012). https://doi.org/10.3151/jact.10.301
23. Goudjil, N., Djelal, C., Vanhove, Y., Kada, H., Heloun, N.: Impact of temperature on the demoulding of concrete elements with a polarization process. Constr. Build. Mater. **54**, 402–412 (2014). https://doi.org/10.1016/j.conbuildmat.2013.12.034

Active Control of 3D Printing of Cementitious Materials

Yi Zhang, Swapnil Balasaheb Ghodke, Dengwu Jiao, Yaxin Tao, Luise Göbel, Shravan Muthukrishnan, Yiyuan Zhang, Paula Heik, Ojas Arun Chaudhari, Akilesh Ramesh, and Geert De Schutter

Abstract 3D concrete printing is widely recognized for its significant potential in resource conservation, environmental protection, the construction of intricate geometries, and operations in harsh environments. Despite these advantages, its performance requirements are stricter and more complex than those of traditional concrete,

Y. Zhang (✉) · Y. Zhang · G. De Schutter
Department of Structural Engineering and Building Materials, Magnel-Vandepitte Laboratory, Ghent University, Ghent, Belgium
e-mail: Yi.Zhang@UGent.be

Y. Zhang
e-mail: Yiyuan.Zhang@UGent.be

G. De Schutter
e-mail: Geert.DeSchutter@UGent.be

S. B. Ghodke
Technology Innovation Center for 3D Printing of Concrete, Department of Civil Engineering, Indian Institute of Technology Roorkee, Roorkee, India
e-mail: swapnilb_ghodke@ce.iitr.ac.in

D. Jiao
Department of Architecture and Civil Engineering, City University of Hong Kong, Kowloon Tong, Hong Kong, China
e-mail: dengwu.jiao@cityu.edu.hk

Y. Tao
Institute for Building Materials, Department of Civil, Environmental, and Geomatic Engineering, ETH Zürich, Zürich, Switzerland
e-mail: yaxin.tao@ifb.baug.ethz.ch

L. Göbel · P. Heik
Chair of Mechanics of Engineering Materials, F. A. Finger-Institute for Building Material Science, Bauhaus-Universität Weimar, Weimar, Germany
e-mail: luise.goebel@uni-weimar.de

P. Heik
e-mail: paula.heik@uni-weimar.de

S. Muthukrishnan
Cluster of Excellence CARE, TU Dresden and RWTH Aachen, Dresden, Germany

G. De Schutter and J. Sanjayan (eds.), *Active Control of Properties of Fresh and Hardening Cementitious Materials*, RILEM State-of-the-Art Reports 50,
https://doi.org/10.1007/978-3-032-30340-0_8

which poses challenge that conventional performance control approaches are often unsuitable and cannot fully exploit the benefits of 3D printing. Therefore, the need to develop intelligent and flexible active performance control strategies for 3D printed cement-based materials becomes increasingly urgent. In this context, we summarize the state of the art in active performance control strategies at different stages of 3D printing. Specifically, in printing stage, pumping, extrusion and deposition were considered, while in post printing stage, we focus on the shrinkage behaviour, interlayer adhesion, mechanical performance and durability. We also provide critical commentary on these strategies, identify current limitations and discuss their prospects. It is expected to offer insights that can guide the development of active performance control in 3D concrete printing and promote its broader application.

Keywords Active performance control · Additive manufacturing · Rheology · Mechanical performance · Durability · Concrete

1 Introduction

Recently, 3D concrete printing (3DCP) or digital concrete technology has attracted growing interest and recognition as a transformative technology in the construction industry. Among various 3DCP techniques, extrusion-based 3DCP received more attention due to the wider application scenario and the relative maturity of material technology, printing equipment, and control technology. Unlike the traditional concrete casting process, extrusion-based 3D concrete printing technology works on digital models by layer-wise extrusion and deposition of materials to target components and buildings.

On the one hand, this technology does not need formwork and automatically deposits on-demand, which saves material, improves construction efficiency, and facilitates personalized customization. Consequently, it has significant advantages in energy saving and environmental protection, construction of intricate geometries, construction in harsh environments, etc., which bring new opportunities for the construction industry [1–3]. On the other hand, due to the innovative processes, new challenges arise with respect to the performance of 3D printed concrete. It requires

Institute of Construction Materials, TUD Dresden University of Technology, Dresden, Germany

S. Muthukrishnan
e-mail: shravan.muthukrishnan@tu-dresden.de

O. A. Chaudhari
RISE Research Institutes of Sweden, Borås, Sweden
e-mail: ojas.arun.chaudhari@ri.se

A. Ramesh
Centre for Future Materials, University of Southern Queensland, Brisbane, Australia
e-mail: aramesh@swin.edu.au

a precise balance of fresh and hardening properties of printed concrete to achieve successful printing, specifically requiring the materials to flow easily and uniformly during pumping and extrusion to ensure continuous and stable properties during printing [4]. Simultaneously, materials should possess the ability to maintain shape stability immediately after extrusion and support the subsequently printed layer. Additionally, the performance of printed structures is highly dependent on the interfacial adhesion between layers, which can be compromised by inadequate rheology control or premature setting [5]. Rapid fabrication and application also require sufficient strength development. Furthermore, the absence of formwork exposes printed material to the environment prematurely, posing challenges related to shrinkage and durability [6, 7].

To address these challenges and meet these performance requirements, active control of the properties of fresh and hardening 3D printed concrete is essential and attracts increasing attention. Previous chapters have reviewed performance-based active control methods for concrete. This chapter delves into the critical role of active control strategies in optimizing extrusion-based concrete 3D printing, highlighting the importance of balancing fresh-state properties and hardening kinetics to ensure print quality, structural integrity, and long-term performance.

2 Performance Requirements at Different Stages

Due to the lack of formwork support, one of the critical challenges of extrusion-based 3DCP lies in the complex and even contradictory property requirements during printing. The evolving set of properties during printing is called printability, including pumpability, extrudability, and buildability. Pumpability refers to the ease of pumping the material from the storage point to the nozzle. This requires materials with good fluidity, i.e., appropriate viscosity and dynamic yield stress, which can reduce the pressure required in the pumping process and improve the conveying efficiency of materials. Additionally, good water retention performance is necessary to avoid the plugging of concrete. Extrudability refers to the ease of extruding the material during the extrusion process and the continuity and surface quality after extrusion. On the basis of the performance requirement during pumping, proper rheological behaviour of the fresh material is necessary to ensure print quality and avoid plastic cracking and interruption of printed materials [4]. Whereas buildability refers to the degree of deformation and overall stability of the 3D printed concrete after extrusion under its self-weight, extrusion pressure, and the gravity force of subsequent layers. This requires the material to possess an adequate yield stress that keeps its shape stable after extrusion [5]. Especially, if there is an area variation in the cross-section of the nozzle, a thixotropic requirement of fast recovery of internal structure is necessary [6, 8]. A suitable structuration rate or setting rate is also important for the layer thickness and printing rate during deposition. Additionally, the period of duration (open time) during which the material can meet the basic pumpability, extrudability, and buildability requirements simultaneously is also a key point for printing. The open time is generally dominated by the setting time of concrete.

Without formwork protection, printed materials are directly exposed to the environment, which enhances water loss and leads to a higher risk of plastic shrinkage cracking. In addition, the higher binder-to-aggregate ratio in the printed concrete raises the risk of autogenous shrinkage and creep. Moreover, the poor adhesion between layers can accelerate the water loss of the interface region and internal printed materials [9]. To mitigate shrinkage cracking, it demands that printable materials should have a good water retention performance or a suitable water loss rate during hardening. Good adhesion between layers will contribute to better volume stability of printed samples.

The layer interface structure is introduced into printed materials and structures owing to the absence of vibration, affecting their structural integrity. The structural integrity is primarily governed by interlayer adhesion performance, which is influenced by the rheology of materials, printing parameters, and the degree of hydration at the layer interface. Poor interlayer adhesion results in porous and weak interfaces between layers, causing anisotropic mechanical and transport properties. In this case, adjustments of rheological properties and hydration of materials, and printing parameters are necessary to reduce the differences in the phase composition and distribution between the interface and the bulk materials.

Apart from the differences in processing techniques, the composition of concrete used for 3DCP is also different compared to traditional concrete construction. These differences critically influence the performance of hardened concrete, including strength development and durability. As a result, the performance control methods for printed concrete require optimization rather than direct application of conventional approaches. Moreover, 3DCP demonstrates substantial potential in special applications, particularly in challenging environments, like tunnel linings, underwater structures, polar regions, and lunar construction. These scenarios impose specific performance requirements, including rapid early-stage strength development, high washout resistance, and exceptional freeze-thaw resistance.

3 Active Control Strategies During Printing

3.1 Active Control Strategies During Pumping

Maintaining proper flowability and stability during pumping is essential for ensuring continuous extrusion and deposition in 3DCP. Currently, two main strategies are employed for the active control of pumpability. One is the active vibration intervention, which can reduce the viscosity and yield stress of concrete by breaking down the internal structures in fresh concrete. The other is active control of the efficacy of admixtures, which are responsible for external triggers, like temperature, magnetic and electric fields, etc., to adjust the rheological properties of the concrete in real time [10–12].

In the context of 3D printing, the printable concrete usually possesses high thixotropy, i.e., its viscosity and yield stress decrease under shearing, while increasing during rest. When vibration is introduced, the mechanical shearing disrupts the flocculated structure formed by cement particles, leading to a significant reduction in yield stress and viscosity. This reduction allows for smoother and more consistent flow, making the material easier to pump [13]. Consequently, vibration is commonly adopted to actively control the flowability during pumping in 3DCP. Normally, the vibration is applied before pumping so that the viscosity and yield stress are lowered and suitable for pumping, ensuring a stable material flow. However, with the pumping distance increasing, the recovery of the internal structures leads to increasing flow resistance. To address this, magnetic-triggering internal vibration, a promising real-time control strategy, was proposed by De Schutter under the framework ERC advanced project 'SmartCast' [11]. By incorporating magneto-responsive particles into the concrete mix, an applied external magnetic field, either constant or alternating, can influence their movement and alignment along magnetic induction lines. In the pumping stage, a time-varying magnetic field generated by an alternating coil has proven effective for inducing internal vibrations. Research by Zhang et al. [14] demonstrated that applying high-frequency magnetic fields of varying intensities can agitate magneto-responsive particles, disrupting flocculation networks within the cement paste and thereby preserving or even restoring the material's fluidity, as shown in Fig. 1.

Despite the potential of this technique, several challenges remain in scaling it up for practical applications. One key limitation is the high cost of magneto-responsive particles. While larger and more affordable magnetite particles could serve as an alternative, their increased weight may lead to segregation or higher pumping pressure requirements. Another challenge lies in achieving a uniform magnetic field distribution throughout the concrete. The field strength tends to decrease from the pipe's outer walls toward the centre, potentially resulting in non-uniform rheological properties. Further research is needed to address these challenges, refine these advanced techniques, and ensure their feasibility for large-scale 3DCP applications.

Zhang et al. [15] reported that increased temperature enhances the molecular chain stretching of polycarboxylate ether-based superplasticizer and their adsorption on cement particles, causing the reduction of viscosity and yield stress of printable materials. De Schutter et al. [12] developed an innovative electromagnetic responsible superplasticizer and demonstrated its ability for active rheology control. Although these efforts offer valuable inspiration for active rheology control during pumping, further investigation is still required for application in 3DCP. More details of these approaches can be found in Chap. "Active Control of Rheology of Cementitious Materials".

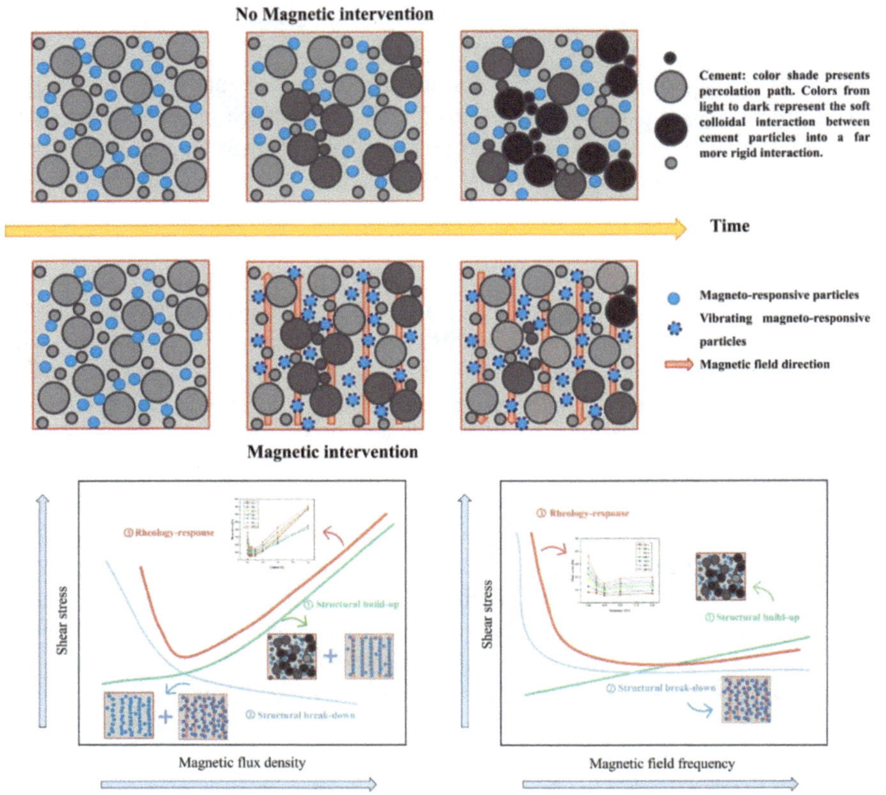

Fig. 1 Active rheology control utilizing an external magnetic field: (top) particle distribution before and after the application of a bipolar magnetic field, (bottom-left) influence of magnetic field strength on shear stress, and (bottom-right) impact of magnetic field frequency on shear stress. Modified from [14]

3.2 Active Control Strategies During Extrusion

Since shape stability after extrusion is crucial for achieving successful 3DCP, the extrusion stage is where active control of the rheological properties is most frequently implemented. This is facilitated by multiple interventions at the print head, making it a key focus in 3DCP compared to other stages. Among them, in-line activation that allows for the local activation of small batches of concrete material and is already established [11]. In this technique, a base mix, characterized by a slow setting rate and high pumpability, is transported to the nozzle for printing. Before deposition, the base mix is combined with chemical admixtures to achieve the desired rheological properties during extrusion. These admixtures may include viscosity-modifying agents and accelerators, depending on the requirements of the printed object and environmental conditions. Detailed explanations of the types of chemical admixtures utilized to enhance the rheological properties of cementitious mixes can be found in [16], while [17, 18] provide detailed insights specific to alkali-activated materials.

A significant limitation of this approach is the design and efficiency of the in-line mixer. This mixer must not only ensure uniform distribution of the chemical admixtures within the base mix but also achieve this within a minimal time frame. The longer the base mix resides in the in-line mixer, the larger the required mixer volume becomes, posing a challenge to the design freedom of 3D printers. In-line mixers are broadly categorized into static and dynamic mixers [19, 20] (Fig. 2). Static mixers function without the need for external energy, relying solely on the flow division pattern of the internal mixing baffles [21–23], whereas dynamic mixers require a motor and electrical connections, which can introduce additional complexity to the print head design [24]. While static mixers provide a simpler and more energy-efficient solution, their mixing efficiency can be constrained by limitations in pumping pressure. Consequently, achieving effective active rheology control in this technique necessitates a coordinated approach, where both mixture design and print head design are optimized to function in synergy.

It is well-known that higher temperatures directly influence reaction rates, such as cement hydration. Therefore, it is straightforward to apply sources of heating at print heads to actively control the setting of concrete, as it is also widely applied in pre-cast industries [25]. Several heating methods exist, e.g., using coils or steam. Still, challenges regarding the non-uniform temperature development in the bulk material exist, with the temperature being highest near the heating elements and lowest at the core, limiting the applicability in 3DCP due to uncontrolled evolutions of the yield strength. Several authors investigated the effects of microwave heating on the buildability of different concrete types [15–17]. Microwave-assisted heating offers more uniform heating. Researchers have shown that microwave heating for just a few seconds can enhance yield strength development and aid in the on-demand setting of concrete. Other methods, such as post-print heating with heat guns or

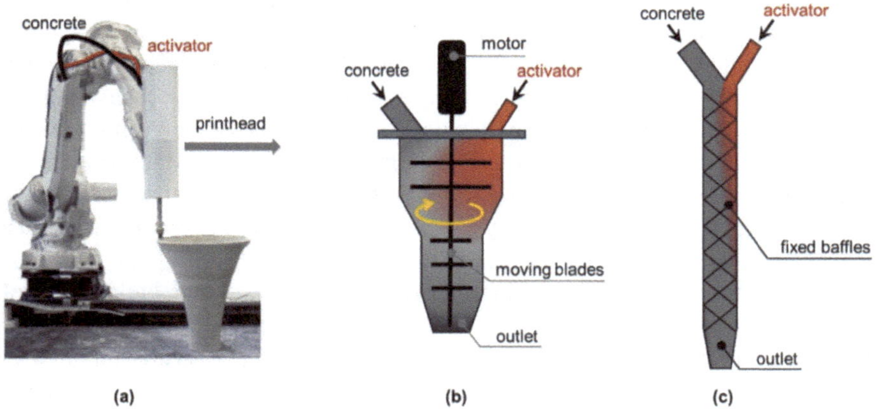

Fig. 2 In-line mixer for active rheology control during extrusion: **a** schematic of printhead on a robot arm; **b** schematic of a dynamic mixing system; **c** schematic of a static mixing system [20]

lamps, have been studied to improve shape stability and buildability. However, they can result in non-uniform heating, limiting their effectiveness. To reduce reliance on the mixing efficiency of in-line mixers, researchers have proposed an alternative approach involving the encapsulation of chemical admixtures. These encapsulated admixtures are introduced during the initial mixing stage of the concrete, where the mixing duration and the print head volume are decoupled. This ensures that the encapsulated accelerators are uniformly distributed within the concrete. The blended concrete is then pumped to the print head, where the capsules are triggered to release the accelerators. Kanagasuntharam et al. [26] encapsulated set accelerators using thermally responsive polymers. These capsules were introduced during the initial mixing phase and later activated thermally at the print head at a constant temperature of 65 °C, such that the capsules melted and released their ingredients. Three different accelerators were used, namely a commercial shotcrete accelerator, NaOH, and Na_2SiO_3. For the capsule wall materials, both gelatine and a thermo-responsive paraffinic phase change material were selected. By means of mini-slump cone and slow penetration test methods, a good pumpability and an increased static yield stress to varying degrees were reported. Noteworthy, the buildability of a demonstrator object (3D printed concrete wall) was significantly improved. However, the impact of the capsules on the mechanical properties was inconsistent, with some mixtures exhibiting improved strength while others showed reductions. Shao et al. [27] used the wall material polyethylene glycol to encapsulate an accelerator and a viscosity modifier. Graphite powder served as a microwave responsive medium. Upon microwave-triggered opening of the capsules, the rheological properties of cement pastes were modified, whereby the extrudability of cement pastes with high flowability was particularly improved.

Another approach of coupling temperature control and viscosity modifiers was studied by Zhang et al. [10, 28]. Rheological experiments at temperatures in a range between 5 and 45 °C were executed, including investigating flow curves, stress growth tests, and small-amplitude oscillatory shear tests. Further printing tests were performed for validation purposes. The findings of this study suggest that increasing the temperature enhances the effectiveness of superplasticizer and viscosity modifier by accelerating their dissolution and adsorption while promoting chain stretching. As a result, temperature variations can either amplify or suppress changes in the apparent viscosity, thixotropic area, static yield stress, and storage modulus of 3D printed concrete. The printing test confirmed that buildability improved with increasing temperatures, although the effect on the extrudability of mixes was negligible for mixes containing a viscosity modifier. This admixture appeared to reduce the temperature dependence of the viscosity of the printed material.

Short-pulsed magnetic fields during extrusion improved the buildability of printed mixes, as explained in Chap. "Active Control of Rheology of Cementitious Materials". The introduction of magnetic forces into 3DCP was first reported in 2016 by Manninger et al. [29]. The authors identified clogging of the extruder as a major challenge in the process. In their study, Fe_3O_4 was added to the concrete mixture, and a magnetic force field was applied using a standard industrial magnet, leading to immediate improvements in the stability of the printed concrete layers.

The mechanism behind using the concept of MR fluids at the print head during 3DCP is illustrated in Fig. 3 (left). Kanagasuntharam et al. [30] used carbonyl iron particles (CIP) as magneto-rheological additives to enhance the buildability of 3DCP. The authors categorized the buildability enhancement methods into (i) print head and (ii) print-bed activation methods. The authors used permanent magnets attached to a frame, enabling a rather low magnetic field of 0.1 T. In their study, the materials were exposed to the magnetic field for 60 s at the print head, but this resulted in only a negligible increase in static yield strength. This observation was attributed to the high percolation time in the mixtures containing magnetic particles and superplasticizer. Besides, Jiao et al. [31] used rheometric tests to simulate the extrusion process by constant shearing and the layer building after extrusion by measuring the structural evolution using small-amplitude oscillatory shear (SAOS) tests. It was found that applying a short pulsed magnetic field during the extrusion could enhance the buildability of the printed materials containing magnetic particles. Considering the high cost of the magneto-responsive materials, Yue et al. [32] suggested the use of electric arc furnace slag powder (EAFSP) as a magneto-responsive additive. The authors reported an increased storage modulus (by a factor of ten or more) when the cement pastes containing EAFSP are exposed to magnetic fields of different strengths. However, it was also acknowledged that the duration at which the printing material is subjected to the magnetic field during the printing process is rather short, see Fig. 3 for a prospective configuration. Therefore, the material responses for exposure to a magnetic field for ten seconds during rheological plate-plate measurements were evaluated and found to be rather insensitive to the magnetic field strength when varied between 0.2 and 0.8 T. Moreover, the buildability of mixes with different concentrations of EAFSP was compared with and without a magnetic field of 0.2 T. To test buildability, a ring-shaped permanent magnet was applied in a cylindrical mould. After removing the mould, the height of the concrete cylinder was measured after a time instant of 30 s. An improved buildability was observed and attributed to the chain-like structures of the responsive particles due to magnetization. Zhang et al. [33] achieved active rheology control by using magneto-responsive aggregates under a nozzle/pipe-like (inline) external magnetic field. An experimental validation, qualitative analysis and quantitative characterization of magnetorheological response and mechanism of cementitious materials containing responsive aggregates were conducted. The developed magneto-responsive mortar exhibited a significant magnetorheological response under the influence of an inline magnetic field. It should be noted that the above methods can induce significant magnetorheological responses, such as increased yield stress and storage modulus, under the influence of a magnetic field. However, due to the reversible magnetization of soft magnetic materials, these properties rapidly decrease once the magnetic field is removed. Recently, to enhance the residual magnetic response particle microstructure and their effects on the rheological properties of materials after magnetic field intervention, Zhang et al. [15] introduced hard magnetic particles (neodymium iron boron, $Nd_2Fe_{14}B$). This innovative approach showed an increase in yield strength and magnetization after magnetic field intervention severalfold. Moreover, the increased rheological properties and magnetization were observed to sustain even after the removal of

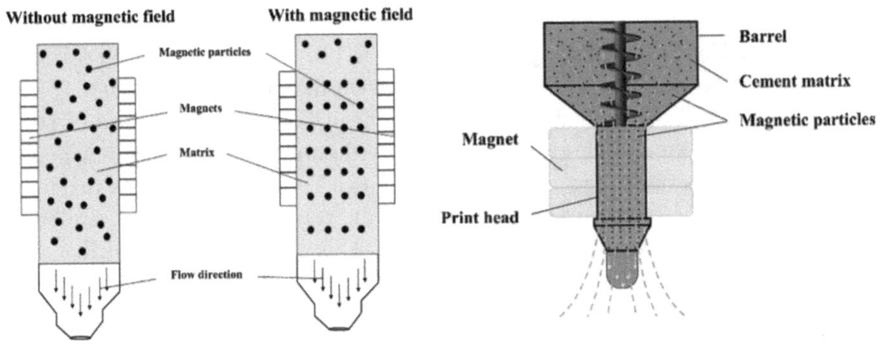

Fig. 3 Left—magnetorheological activation applying the concept of MF fluids at the print head during the extrusion process (from [16]) and concept for applying a magnetic source at the print head (from [32])

the magnetic field. This sustained improvement in the yield strength guarantees an improved buildability for 3DCP.

Another intervention technique is referred to ultrasonication at the print head (Fig. 4). Vaitkevičius et al. [34] aimed for rapid setting of concrete using ultrasonic pulses by connecting an ultrasound dispersion machine to a customized 3D printer. The authors studied the mechanical properties of the freshly printed mixes and were able to achieve a compressive strength of 1 MPa after 20 min, while no measurable strength values could be obtained without ultrasonic activation.

The application of external vibration effects to actively control the rheological properties of printable concrete was explored by Sanjayan et al. [13]. The authors examined the effects of vibration frequency and acceleration with a particular focus on reducing the extrusion pressure. It was found that the extrusion pressure and yield stress are not clearly affected by the peak velocity of the vibration until it reaches a critical value. Beyond this point, referred to as the critical vibrational peak velocity value, both parameters begin to decrease. Further, evaluating the material mix properties showed that very low-viscous fluids and stiff mixes exhibit a lower vibrational peak value. Instead of directly enhancing buildability, the introduced vibration improves flowability, thereby enhancing extrusion quality. This suggests that the printable material is initially quite stiff before extrusion. However, this approach may present challenges for long-distance pumping, where maintaining optimal flowability is crucial.

3.3 *Active Control Strategies During Deposition*

Unlike the extrusion stage, active control strategies for the deposition stage refer to approaches applied to the materials once extruded (at the print bed) to enhance

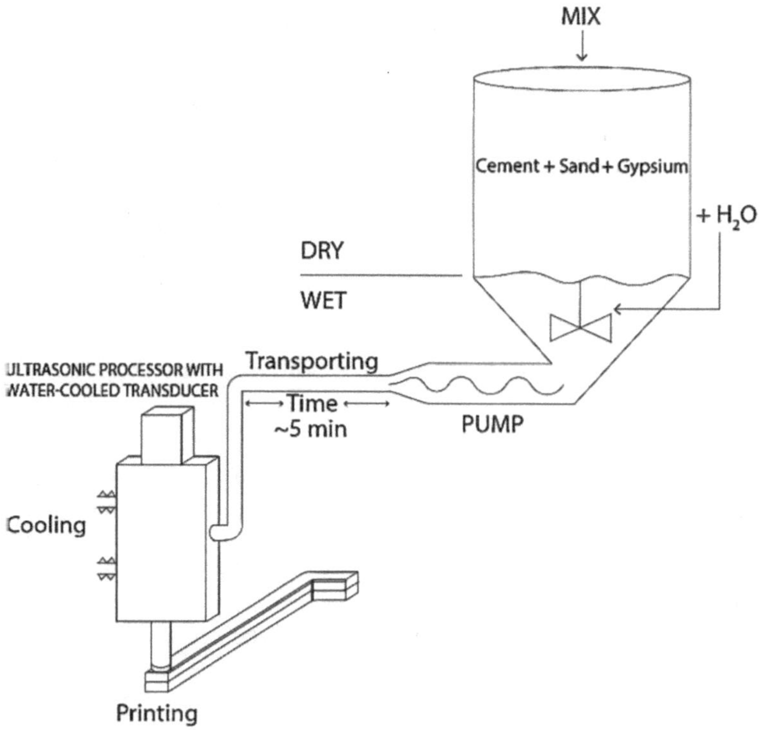

Fig. 4 Application of an ultrasonic dispersion machine within a 3DCP process [34]

buildability, further improving printing efficiency in the layer-up-layer process. So far, relevant research remains limited, primarily due to challenges associated with upscaling and automation. For instance, modifications made to the print head to achieve active rheology control during the extrusion process are more easily integrated into the printer and require minimal changes when printing larger structures. However, implementing control strategies at the print bed necessitates additional systems and integrations to enable automation, as well as significant adjustments to accommodate varying print sizes.

Despite these challenges, previous studies have explored promising approaches to enhance the setting rate of concrete during deposition, as it is generally undesirable to reduce the setting rate. In a study by Battacherjee et al. [35], an accelerator was sprayed onto the filaments post-extrusion to enhance the setting rate of the concrete. This method eliminated the need for a print head mixer to achieve improved buildability using the accelerator. However, spraying the accelerator may lead to a non-uniform setting of the printed layers, with the outer surface setting faster than the core. In a similar study, Tay et al. [36] sprayed CO_2 with steam to accelerate the setting rate of the printed layer (as shown in Fig. 5). The use of CO_2 as an accelerant

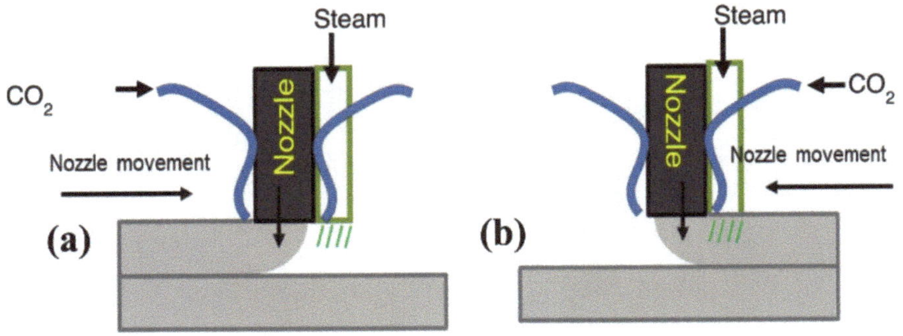

Fig. 5 Illustration of spraying CO_2: **a** separately and **b** with steam on the printed layers. The image is reproduced from [36]

also offers significant environmental advantages. Similar to the spraying of commercial accelerants, this method resulted in a non-uniform setting, with the outer surface of the printed layers experiencing a faster setting rate.

Regarding heat as an intervention for setting control, Kazemian et al. [37] utilized a hot air gun operating at 1200 W and 528 °C post-extrusion of concrete to accelerate the hydration reaction and promote faster setting. The printed layer subjected to heating exhibited a deformation of 0.6 mm under load, in contrast to 2.1 mm of deformation in the layers that were not subjected to heating. Similarly, Dai et al. [38] enhanced the stiffing behaviour and buildability of fly ash-based geopolymer through the hot air gun method.

In all the studies mentioned above, whether investigating the application of accelerators, CO_2, or heat, the setting rate of the outer surface was found to be faster than that of the core. In contrast, Kanagasuntharam et al. [30] applied print-bed magnetization to a magnetorheological cement mortar (MRC) mixture containing carbonyl iron particles (CIP) to actively control its rheological properties. Their findings revealed that continuous magnetization at the print bed significantly increased static yield stress compared to brief magnetization at the print head. Additionally, compressive strength improved with CIP content up to 5%, but higher dosages led to a decline in mechanical performance.

4 Active Control After Printing

4.1 Active Control Strategies for Shrinkage Behaviour

Existing methods for active shrinkage control predominantly rely on regulating external environmental factors, such as humidity, temperature, and gas atmosphere (e.g., CO_2). However, these approaches face challenges in achieving precise shrinkage control and often cause high costs and waste, limiting their practical applicability.

Concrete shrinkage is closely linked to its water content and the rate of water loss. The key factors influencing this phenomenon include the water-to-binder ratio and the water retention capability of concrete, along with environmental conditions such as humidity, wind speed and temperature [39, 40]. To mitigate autogenous shrinkage, actively compensating for the consumed water is an effective strategy [39, 41]. The active control strategies for plastic and dry shrinkage behaviour can be divided into two main categories. One of which is to prevent the water loss of concrete, e.g., covering its surface with foil or a shrinkage-reducing agent [42]. The second method is to compensate for the water loss of concrete, e.g., moistening the sample with water [43].

Moelich et al. [44] investigated the influences of water spraying and curing agents on the shrinkage and cracking behaviour of printed samples. A 19% reduction of the 3 h shrinkage strain and a 70% crack area reduction were achieved when they implemented multiple water sprays at 20 min intervals. Additionally, following each water application, the shrinkage rate dropped to nearly zero for 9 to 12 min and then returns to that of the reference. They found that the use of curing agents on the printed sample surfaces mitigates pore water evaporation, hence remarkably reducing the 3 h free shrinkage by 34%. The curing agent (CA) delays cracking by about 30 min and finally reduces the crack area by 66%.

Bekaert et al. [45] found that reducing the duration of curing in a low RH environment before water compensation is beneficial for a decrease in the shrinkage strain of printed samples. Additionally, applying external moisture on printed concrete that is cured at lower RH for longer periods still results in shrinkage compensation or even swelling of samples.

Liu et al. [46] explored the effects of carbonation mixing as well as carbonation curing on the shrinkage behaviour of printed samples (Fig. 6). Their findings show that compared to the reference, carbonation mixing and carbonation curing reduced the shrinkage rate in the horizontal orientation by 14% and 77%, respectively. Moreover, they pointed out that carbonation accelerates the shrinkage development of samples, enabling their shrinkage deformation to reach stable stages more quickly. They attributed this phenomenon to the formation of dense carbonates, which fill the pores of samples and inhibit water evaporation.

4.2 Active Control Strategies for Interlayer Adhesion

The hardened properties of 3D printed concrete depend on the interface quality, which further relies on the phase distribution, including the pores, aggregates and hydrates at the layer interface [47, 48]. The phase distribution is influenced by various factors, such as environmental conditions, process parameters, rheological properties and material reactions [49–52]. Consequently, these factors can be actively

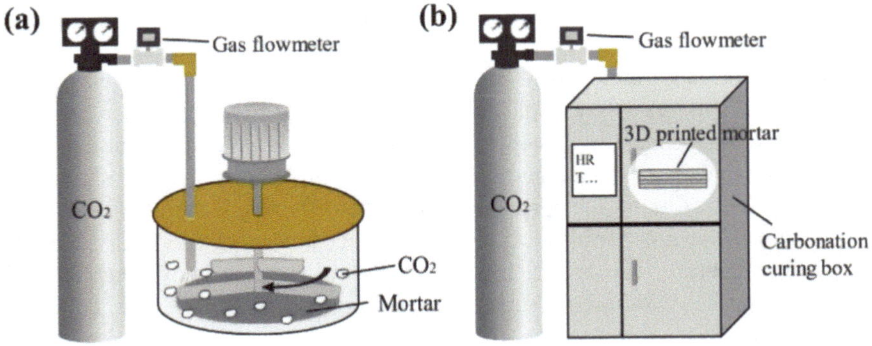

Fig. 6 Illustration of carbonation technologies for printed concrete: **a** carbonation mixing, **b** carbonation curing. Modified from [46]

controlled to enhance the homogeneous phase distribution at the layer interface, therefore improving the interlayer adhesion.

Chen et al. [53] used four different techniques, such as air curing, plastic film covering, wet towel covering and water mist 30–60 min for long time intervals (4 h) between the printing sessions to assess the interlayer bond strength. The results indicate that the plastic film covering and a wet towel covering before placing the second layer significantly enhance the interlayer bond strength as compared to the exposing layer in an ambient environment. They attributed this to the slower surface drying rate during resting, which may decrease the small air void (<25 μm) content, capillary porosity, and micro-cracks localized at the interface. However, applying water mist every 30 or 60 min was found to have detrimental effects on interlayer bonding. A layer of portlandite and/or calcite was observed at the interface of printed samples treated with water mist, which may weaken the interlayer adhesion. In addition, Li et al. [54] studied the effect of various curing conditions, including air curing, steam curing, constant standard curing, and CO_2 curing, on the interlayer bond strength. They found that the air curing or steam curing resulted in a higher porosity and loose microstructure. This resulted in decreased interlayer bond strength after air curing. However, steam curing resulted in an initial increase in the interlayer bond strength followed by a decrease in later-age bond strength. Whereas CO_2 curing enhances the performance of the micro-interface and refines the pore structure. Simultaneously, CO_2 curing demonstrates the capability to improve interlayer bond strength and mitigate strength degradation induced by low humidity without altering the rheological properties of the new paste.

Alternatively, Zhang et al. [55] focused on the post-printing temperature influences on interlayer adhesion of printed samples. They found that increasing the post-printing temperature reduced the discrepancies in the aggregate volume fraction and porosity between the interface and the bulk matrix. Furthermore, a more concentrated distribution of atomic ratios and elastic modulus of hydrates was observed at the layer interface after curing at higher temperatures. As a result of the improved phase distribution at the layer interface, the interlayer bond strength was enhanced. Furthermore, Munemo et al. [56] developed a novel method in which thermo-hydrokinetics was used to enhance the interlayer bond strength. It was produced by spraying steam on the printed layer before placing the next layer (See Fig. 7). The method was implemented for three distinct passing durations: the typical pass time for the arbitrarily chosen object for printing, as well as 5 and 10 min. A control for each pass time, on which thermo-hydrokinetics were not activated, is also collected. The results indicate that the application of thermo-hydrokinetics enhances interlayer bond strength, even over an extended duration.

Inspired by the construction of brick houses, an extra bonding layer was explored to enhance the interlayer adhesion. Hosseini et al. [57] introduced a polymer layer composed of sulphur and black carbon at the interface between adjacent layers, achieving an improvement of over 100% in interlayer bond strength. Similarly, Marchment et al. [58] use cement paste as a bonding agent. Prior to applying the upper layer, a cement paste with sustained flow properties is deposited above the lower layer to minimize voids and enhance the bonding area (See Fig. 8). The results indicate a higher bond strength as compared to the samples with paste. Subsequent investigation indicates that the bond area in the samples with applied pastes is significantly influenced by stiffening effects and surface moisture levels. Recently, Teng et al. [59] utilized strain-hardening cementitious composites (SHCC) as the bonding layer to simultaneously enhance the interlayer adhesion and flexural performance of 3D printed concrete. They reported that SHCC bonding layer enhanced the interlayer bond strength by about 80% and reduced the interfacial porosity by about 35%. Additionally, the flexural strength, deflection and energy absorption of the printed beams were improved by 25%, 180% and 800%, respectively. Even though the above methods proved effective in enhancing interlayer bonding performance, concerns remain regarding the effect from long-term aging of the bonding layer and potential volume instability due to the differing composition from the printed layers.

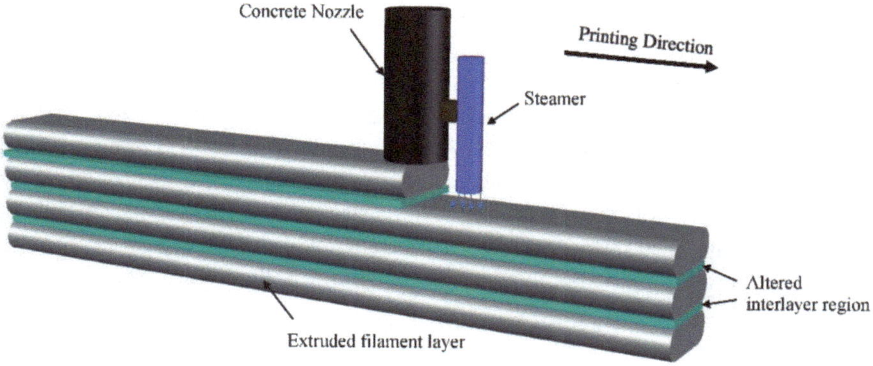

Fig. 7 Schematic of printing setup for interlayer adhesion active control through steam treatment [56]

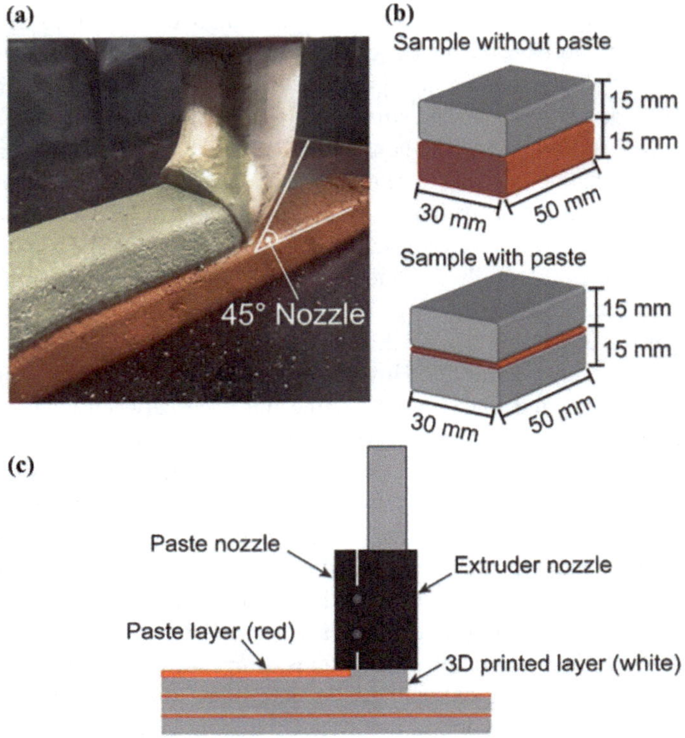

Fig. 8 **a** Mortar mix extruded from a 45° angle nozzle without an interlayer paste mixture, **b** Samples with and without an interlayer paste mixture placed between layers. **c** A diagram of the proposed dual nozzle extruder applying the paste layer and the 3D printed layer [58]

Several researchers have also explored the active control of interlayer adhesion by enhancing the surface roughness of printed layers. He et al. [60] evaluated the impact of a tooth-like interface on the interlayer adhesion of printed objects. For this objective, a tooth angle of 45° was employed (See Fig. 9), increasing interlayer tensile strength and shear strength by 294% and 89%, respectively. The authors noted a transition in the failure mechanism from exclusive adhesive failure to a combination of adhesive and cohesive failures. Similarly, Van der Putten et al. [61] also used various techniques like wire brushing, the addition of sand or cement and moisturizing the substrate layer to increase the surface roughness of printed layers and observed their impact on the interlayer bond strength for 0 and 30 min interlayer printing time gaps. The surface roughness generated by the comb was observed to be effective in increasing the interlayer bond strength for both interlayer printing time intervals.

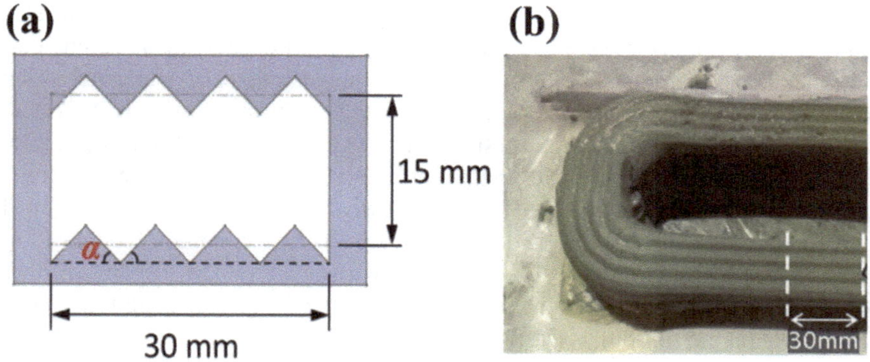

Fig. 9 **a** Layout of the tooth-like nozzle. **b** Printed layer surface [60]

4.3 *Active Control Strategies for Other Performances During Service Life*

Active control of mechanical strength

The mechanical strength of concrete is primarily determined by the degree of cement hydration and the presence of defects. In the case of 3D printed concrete, defects are mainly dominated by layer adhesion and shrinkage-induced cracking [43, 55]. Consequently, the active control strategies for enhancing interlayer adhesion and mitigating shrinkage are also efficient for mechanical strength control. Relevant aspects are discussed in Sects. 4.1 and 4.2 and will not be repeated here. Beyond these, optimized curing conditions, like tailored temperature and humidity, are commonly adopted for strength control of 3D printed concrete structures by enhancing both the degree and uniformity of cement hydration [62].

Wei et al. [63] examined the effects of ambient temperature and humidity on the mechanical properties of 3D printed concrete. They found that setting the curing temperature at 20 ± 1 °C and maintaining a relative humidity of 95% enhanced the material's mechanical properties. Conversely, low temperatures and humidity levels adversely affected the concrete's strength. Sun et al. [64] also found that the development of compressive strength of printed concrete accelerated with the increase of curing temperature, due to the acceleration of cement hydration. However, the anisotropic behaviour is less significant with increasing the curing temperature. Alchaar et al. [65] observed that the flexural strength of 3D printed fibre-reinforced concrete cured at high temperature (40–45 °C) increased by 18% compared to that of specimens curing at 24 ± 2 °C. They observed the improved fibre orientation in specimens cured at high temperature and attributed it to the decreased viscosity.

Wang et al. [55] investigated the impact of steam curing on the mechanical performance of 3D printed concrete. A temperature rise rate of 15 °C/h, a sustained temperature time of 10 h, and a sustained temperature of 60 °C were identified as the optimal curing conditions for compressive strength of 3D printed concrete. Furthermore, Yao et al. [56] found that steam curing is significantly beneficial to early strength but leads to the reduction of late strength. As the temperature rise rate and sustained temperature time increased, the anisotropy of 3D printed samples enhanced, while increasing the sustained temperature leads to a reduction in the anisotropy. Similarly, Sun et al. [57] reported that low temperatures (10 °C) and high humidity (100% RH and 80% RH) caused increased anisotropies in the mechanical properties of printed concrete. By contrast, Ma et al. [66] found that the compressive and flexural strength of printed concrete decreased by 20–30% under 60% ± 5% RH (dry condition) compared to standard condition (RH > 95%). Besides, wind with an air velocity of 3 m/s further slows down strength development, but the drying and wind curing conditions had a slight effect on the splitting tensile strength. Recently, Bawarith et al. [58] explored the integration of carbon dioxide and steam during the 3D printing process. This approach not only enhances the mechanical properties of the concrete but also offers a sustainable solution by in situ capturing of CO_2. The combined use of CO_2 and steam has been shown to improve the curing process, leading to stronger and more durable structures. In addition to the temperature and humidity, some other trigger signals can also have a significant influence on the mechanical properties of 3D printed concrete. Vaitkevičius et al. [34] found that by attaching an ultrasonic dispersion apparatus at the extrusion nozzle, the mechanical properties of printed concrete can be increased by 10%.

Active control of durability

The active control strategies for shrinkage reduction and interlayer adhesion improvement contribute to the initial structural integrity and performance, hence beneficial to the durability of 3D printed concrete. To improve the impermeability, Nienaber et al. [67] applied silicones both on the top surface of each printed layer and on the side surfaces of hardened 3D printed concrete. Their findings indicated that both methods reduced water absorption, with the application of silicones between layers proving more efficient. However, this approach also reduced the interlayer bond strength.

Merely enhancing initial material properties is insufficient for active control of durability. Ensuring long-term performance throughout the service life of 3D printed concrete is also a crucial aspect of active durability control. Up to date, the active control strategies specifically addressing durability throughout the service life of 3D printed concrete have hardly been reported. Since the importance of active durability control in reducing maintenance costs and ensuring structural safety, research in this field is expected to gain increasing attention in the future. Some active control strategies for the durability of traditional concrete can provide some inspiration, like graded pore structure design [68], real-time monitoring [69], magnetic-driven crack filling [70], mini-vascular network, stimuli-responsive capsules and memory alloys for self-healing [71–74].

5 Summary and Prospects

This chapter provided an in-depth and comprehensive discussion of active control techniques used during and after the 3D concrete printing process. To address different performance requirements, various active control approaches were proposed and investigated, especially at different stages of printing. During the pumping stage, active rheology control through vibration and magnetic intervention was usually adopted to ensure a stable flow without any clogging or segregation. Whereas during the extrusion process, hydration control is more common, which is achieved through the application of accelerators at the print head, heating at the print head, and the use of temperature-dependent encapsulated accelerators. These methods usually efficiently improve buildability after extrusion, while the homogeneity still needs to improve and further attention to be paid to the interlayer adhesion. Compared to these strategies, some focus on rheology control may have lower risks of uneven strength development and weak interlayer adhesion. However, magnetic intervention on materials containing soft magnetic particles showed unideal effects on improving the shape stability after extrusion. Hard magnetic materials were proved to be a better solution for enhancing buildability. During the deposition process, active control techniques such as spraying commercial accelerators, heating the layers, and spraying CO_2 with steam were applied to enhance the buildability. Because these methods act from outside to inside of materials, may enlarge discrepancies between the interface and the bulk materials. A magnetic intervention at the print bed can provide a uniform influence. However, this poses challenges like requiring significant adjustments to match varying print sizes.

Environmental condition control is an effective active control strategy for plastic and dry shrinkage behaviour and mechanical performances of printed concrete. Despite the effectiveness of environmental conditions control, they are weak in precise shrinkage control and cause high costs and waste, especially in-situ curing through modified print heads. Carbonation mixing and curing can achieve active control for shrinkage behaviour. As for interlayer adhesion, active control techniques, such as applying flowable cement paste layers and interlock structures between layers, proved flexible and effective.

Compared to fresh 3D printed concrete, the active control strategies for hardened 3D printed concrete during and after printing are fewer, especially for durability. However, the property of hardened concrete is crucial for the service life of printed structures and buildings. Therefore, applying the various interventions at various stages of the printing process is a promising way to achieve the quality, consistency, and structural integrity of printed elements. In addition, for the active control strategies after extrusion (during deposition and after printing), a fundamental drawback of these methods is their dependence on external conditions rather than intrinsic material properties. Consequently, there remains a critical need for effective internal material regulation strategies to enable more precise and stable performance optimization. Active control approaches began to transform into more cost-effective and sustainable methods, like using by-products as magnetic responsive materials and the utilization of CO_2, which makes it more flexible and closer to practical applications.

References

1. De Schutter, G., Lesage, K., Mechtcherine, V., Nerella, V.N., Habert, G., Agusti-Juan, I.: Vision of 3D printing with concrete—Technical, economic and environmental potentials. Cem. Concr. Res. **112**, 25–36 (2018)
2. Kaliyavaradhan, S.K., Ambily, P.S., Prem, P.R., Ghodke, S.B.: Test methods for 3D printable concrete. Autom. Constr. **142**, 104529 (2022)
3. Perrot, A., Jacquet, Y., Caron, J.F., Mesnil, R., Ducoulombier, N., De Bono, V., Sanjayan, J., Ramakrishnan, S., Kloft, H., Gosslar J.: Snapshot on 3D printing with alternative binders and materials: earth, geopolymers, gypsum and low carbon concrete. **185**, 107651 (2024)
4. Roussel, N.: Rheological requirements for printable concretes. Cem. Concr. Res. **112**, 76–85 (2018)
5. Zhang, Y., Jiang, Z., Zhu, Y., Zhang, J., Ren, Q., Huang, T.: Effects of redispersible polymer powders on the structural build-up of 3D printing cement paste with and without hydroxypropyl methylcellulose. Constr. Build. Mater. **267**, 120551 (2021)
6. Mechtcherine, V., Bos, F.P., Perrot, A., da Silva, W.L., Nerella, V.N., Fataei, S., Wolfs, R., Sonebi, M., Roussel, N.: Extrusion-based additive manufacturing with cement-based materials—Production steps, processes, and their underlying physics: a review. Cem. Concr. Res. **132**, 106037 (2020)
7. Zhang, Y.: Research on rheological properties and interlayer interface characteristics of 3D printed cement-based materials. Ghent University (2024)
8. Ducoulombier, N., Mesnil, R., Carneau, P., Demont, L., Bessaies-Bey, H., Caron, J.-F., Roussel, N.: The "Slugs-test" for extrusion-based additive manufacturing: protocol, analysis and practical limits. Cem. Concr. Compos. **121**, 104074 (2021)
9. Zhu, L., Zhang, M., Zhang, Y., Yao, J., Yang, G., Guan, X., Zhao, Y.: Research progress on shrinkage properties of extruded 3D printed cement-based materials. J. Build. Eng. **77**, 107394 (2023)
10. Zhang, Y., Ren, Q., Dai, X., Tao, Y., Zhang, Y., Jiang, Z., Van Tittelboom, K., De Schutter, G.: A potential active rheology control approach for 3D printable cement-based materials: coupling of temperature and viscosity modifiers. Cem. Concr. Compos. 105496 (2024)
11. De Schutter, G., Lesage, K.: Active Rheology Control of Cementitious Materials. CRC Press, London (2023)
12. De Schutter, G., Ezzat, M., Lesage, K., Hoogenboom, R.: Responsive superplasticizers for active rheology control of cementitious materials. **165**, 107084 (2023)
13. Sanjayan, J.G., Jayathilakage, R., Rajeev, P.: Vibration induced active rheology control for 3D concrete printing. Cem. Concr. Res. **140**, 106293 (2021)
14. Zhang, Y., Tao, Y., Sun, Y., Van Tittelboom, K., Zhang, Y., Lesage, K., De Schutter, G.: Active control of thixotropy of magneto-responsive cementitious materials with the intervention of time-varying magnetic fields. **181**, 107520 (2024)
15. Zhang, Y., Zhang, Y., Tao, Y., Dai, X., Van Tittelboom, K., Lesage, K., De Schutter, G.: Active rheology control of cementitious materials containing hard magnetic particles: Sustained response after magnetic intervention. Cem. Concr. Compos. 106024 (2025)
16. Muthukrishnan, S., Ramakrishnan, S., Sanjayan, J.: Technologies for improving buildability in 3D concrete printing. Cem. Concr. Compos. **122**, 104144 (2021)
17. Muthukrishnan, S., Ramakrishnan, S., Sanjayan, J.: In-line activation of geopolymer slurry for concrete 3D printing. **162**, 107008 (2022)
18. Muthukrishnan, S., Ramakrishnan, S., Sanjayan, J.: Rapid early age strength development of in-line activated geopolymer for concrete 3D printing. Constr. Build. Mater. **406**, 133312 (2023)

19. Tao, Y., Rahul, A.V., Lesage, K., Yuan, Y., Van Tittelboom, K., De Schutter, G.: Stiffening control of cement-based materials using accelerators in inline mixing processes: possibilities and challenges. Cem. Concr. Compos. **119**, 103972 (2021)
20. Wangler, T., Tao, Y., Das, A., Mahmoudi, M., Gürel, S., Flatt, R.J.: Aluminate 2K systems in digital concrete: process, design, chemistry, and outlook. **185**, 107644 (2024)
21. Tao, Y., Lesage, K., Van Tittelboom, K., Yuan, Y., De Schutter, G.: Twin-pipe pumping strategy for stiffening control of 3D printable concrete: from transportation to fabrication. Cem. Concr. Res. **168**, 107137 (2023)
22. Tao, Y., Rahul, A.V., Mohan, M.K., Van Tittelboom, K., Yuan, Y., De Schutter, G.: Blending performance of helical static mixer used for twin-pipe 3D concrete printing. Cem. Concr. Compos. **134**, 104741 (2022)
23. Tao, Y., Mohan, M.K., Rahul, A.V., De Schutter, G., Van Tittelboom, K.: Development of a calcium sulfoaluminate-Portland cement binary system for twin-pipe 3D concrete printing. Cem. Concr. Compos. **138**, 104960 (2023)
24. Reiter, L., Wangler, T., Anton, A., Flatt, R.J.: Setting on demand for digital concrete—Principles, measurements, chemistry, validation. Cem. Concr. Res. **132**, 106047 (2020)
25. Ba, M., Qian, C.: Hydration evolution of pre-cast concrete with steam and water curing. J. Cent. South Univ. **20**(10), 2870–2878 (2013)
26. Kanagasuntharam, S., Ramakrishnan, S., Sanjayan, J.: Active rheology control of concrete using encapsulated accelerator as responsive additives for concrete 3D printing. In: Lowke, D., Freund, N., Böhler, D., et al. (eds.) Proceedings of the 4th RILEM Int'l Conference on Concrete and Digital Fabrication, pp. 244–251. Springer (2024)
27. Shao, L., Feng, P., Zuo, W., Wang, H., Geng, Z., Liu, Q., Miao, C., Liu, Z.: A novel method for improving the printability of cement-based materials: controlling the releasing of capsules containing chemical admixtures. Cem. Concr. Compos. **128**, 104456 (2022)
28. Zhang, Y., Van Tittelboom, K., De Schutter, G., Jiang, Z.: Active rheology control for 3D printable cement-based materials by temperature: an exploratory study. In: Tan, M.J., Li, M., Tay, Y., et al. (eds.) Construction 3D Printing. Springer, pp. 90–97 (2023)
29. Menninger, S., Del Campo, M.: Magnetically controlled concrete 3D printing. In: Kawaguchi, K., Ohsaki, M., Takeuchi, T. (eds.) Proceedings of the IASS Symposium 2016: Spatial Structures in the 21st Century. International Association for Shell and Spatial Structures (IASS), no. 9, pp. 1–9, Tokyo, Japan (2016)
30. Kanagasuntharam, S., Ramakrishnan, S., Muthukrishnan, S., Sanjayan, J.: Effect of magnetorheological additives on the buildability of 3D concrete printing. J. Build. Eng. **74**, 106814 (2023)
31. Jiao, D., Shi, C., De Schutter, G.: Magneto-rheology control in 3D concrete printing: a rheological attempt. Mater. Lett. **309**, 131374 (2022)
32. Yue, H., Zhang, Z., Hua, S., Gao, Y., Cui, S., Wang, Y., Zhang, Z., Zhu, H.: Solid waste-based set-on-demand 3D printed concrete: active rheological control of cement-based magneto-rheological fluids. Constr. Build. Mater. **404**, 133269 (2023)
33. Zhang, Y., Tao, Y., Zhang, Y., Godinho, J.R., Van Tittelboom, K., Lesage, K., De Schutter, G.: Active rheology control of responsive mortar with low slump value: effect of magnetic field properties. Constr. Build. Mater. **458**, 139450 (2025)
34. Vaitkevičius, V., Šerelis, E., Kerševičius, V.: Effect of ultra-sonic activation on early hydration process in 3D concrete printing technology. Constr. Build. Mater. **169**, 354–363 (2018)
35. Bhattacherjee, S., Santhanam, M.: Investigation on the effect of alkali-free aluminium sulfate based accelerator on the fresh properties of 3D printable concrete. Cem. Concr. Compos. **130**, 104521 (2022)
36. Tay, Y.W.D., Lim, S.G., Fadhel, B.A., Amr, I.T., Bamagain, R.A., Al-Hunaidy, A.S., Paul, S.C., Tan, M.J.: Potential of carbon dioxide spraying on the properties of 3D concrete printed structures. Carbon Capture Sci. Technol. **13**, 100256 (2024)
37. Kazemian, A., Yuan, X., Meier, R., Cochran, E., Khoshnevis, B.: Construction-scale 3D printing: shape stability of fresh printing concrete. In: ASME International Conference on

Manufacturing Science and Engineering (ed) Proceedings of the ASME 2017 12th International Manufacturing Science and Engineering Conference, MSEC2017. American Society of Mechanical Engineers (2017)
38. Dai, X., Kandy, S.B., Neithalath, N., Kumar, A., Bauchy, M., Garboczi, E., Gaedt, T., Srivastava, S., Sant, G.: Thermally stimulated stiffening and fly ash's alkaline activation by $Ca(OH)_2$ addition facilitates 3D-printing. Cem. Concr. Compos. **156**, 105870 (2025)
39. Van Der Putten, J., Snoeck, D., de Coensel, R., De Schutter, G., Van Tittelboom, K.: Early age shrinkage phenomena of 3D printed cementitious materials with superabsorbent polymers. J. Build. Eng. **35**, 102059 (2021)
40. Shalaby, Y., Tian, Y., Xu, Y., Yuan, Q.: Advances and challenges in fiber-reinforced shotcrete: a comprehensive review of fresh and placement characteristics. Tunn. Undergr. Space Tech. **166**, 106996 (2025)
41. Tao, J.-L., Lin, C., Luo, Q.-L., Long, W.-J., Zheng, S.-Y., Hong, C.-Y.: Leveraging internal curing effect of fly ash cenosphere for alleviating autogenous shrinkage in 3D printing. Constr. Build. Mater. **346**, 128247 (2022)
42. Sirajuddin, M., Gettu, R.: Plastic shrinkage cracking of concrete incorporating mineral admixtures and its mitigation. Mat. Struct. **51**(2) (2018)
43. Federowicz, K., Kaszyńska, M., Zieliński, A., Hoffmann, M.: Effect of curing methods on shrinkage development in 3D-printed concrete. Materials (Basel, Switzerland) **13**(11), 2590 (2020)
44. Moelich, G.M., Kruger, P.J., Combrinck, R.: Mitigating early age cracking in 3D printed concrete using fibres, superabsorbent polymers, shrinkage reducing admixtures. B-CSA Cement Curing Meas. **159**(106862), 0008–8846 (2022)
45. Bekaert, M., Van Tittelboom, K., De Schutter, G.: Influence of curing conditions on the shrinkage behaviour of three-dimensional printed concrete formwork. Front. Struct. Civ. Eng. **18**(8), 1225–1236 (2024)
46. Liu, Q., Tang, H., Chen, K., Peng, B., Sun, C., Singh, A., Li, J.: Utilizing CO_2 to improve plastic shrinkage and mechanical properties of 3D printed mortar made with recycled fine aggregates. Constr. Build. Mater. **433**, 136546 (2024)
47. Zhang, Y., Ren, Q., Van Tittelboom, K., De Schutter, G., Jiang, Z.: Layer interface in 3D printed cement-based materials: heterogeneous phase distribution and new insights into formation mechanism. Cem. Concr. Compos.
48. Kloft, H., Krauss, H.-W., Hack, N., Herrmann, E., Neudecker, S., Varady, P.A., Lowke, D.: Influence of process parameters on the interlayer bond strength of concrete elements additive manufactured by Shotcrete 3D printing (SC3DP). Cem. Concr. Res. **134**, 106078 (2020)
49. Sanjayan, J.G., Nematollahi, B., Xia, M., Marchment, T.: Effect of surface moisture on interlayer strength of 3D printed concrete. Constr. Build. Mater. **172**, 468–475 (2018)
50. Keita, E., Bessaies-Bey, H., Zuo, W., Belin, P., Roussel, N.: Weak bond strength between successive layers in extrusion-based additive manufacturing: measurement and physical origin. Cem. Concr. Res. **123**, 105787 (2019)
51. Zhang, Y., Zhu, Y., Ren, Q., He, B., Jiang, Z., Van Tittelboom, K., De Schutter, G.: Comparison of printability and mechanical properties of rigid and flexible fiber-reinforced 3D printed cement-based materials. Constr. Build. Mater. **400**, 132750 (2023)
52. Yao, H., Xie, Z., Li, Z., Huang, C., Yuan, Q., Zheng, X.: The relationship between the rheological behaviour and interlayer bonding properties of 3D printing cementitious materials with the addition of attapulgite. Constr. Build. Mater. **316**, 125809 (2022)
53. Chen, Y., Chang, Z., He, S., Çopuroğlu, O., Šavija, B., Schlangen, E.: Effect of curing methods during a long time gap between two printing sessions on the interlayer bonding of 3D printed cementitious materials. Constr. Build. Mater. **332**, 127394 (2022)
54. Li, Q., Gao, X., Su, A., Lu, X.: Interlayer adhesion strength of 3D-printed cement-based materials exposed to varying curing conditions. J. Build. Eng. **74**, 106825 (2023)
55. Zhang, Y., Tao, Y., Godinho, J.R., Ren, Q., Jiang, Z., Van Tittelboom, K., De Schutter, G.: Layer interface characteristics and adhesion of 3D printed cement-based materials exposed to post-printing temperature disturbance. Cem. Concr. Compos. **155**, 105837 (2025)

56. Munemo, R., Kruger, J., van Zijl, G.P.: Improving interlayer bond in 3D printed concrete through induced thermo-hydrokinetics. Constr. Build. Mater. **393**, 132121 (2023)
57. Hosseini, E., Zakertabrizi, M., Korayem, A.H., Xu, G.: A novel method to enhance the interlayer bonding of 3D printing concrete: an experimental and computational investigation. Cem. Concr. Compos. **99**, 112–119 (2019)
58. Marchment, T., Sanjayan, J., Xia, M.: Method of enhancing interlayer bond strength in construction scale 3D printing with mortar by effective bond area amplification. Mater. Des. **169**, 107684 (2019)
59. Teng, F., Ye, J., Yu, J., Li, H., Weng, Y., Mechtcherine, V.: Development of strain-hardening cementitious composites (SHCC) as bonding materials to enhance interlayer and flexural performance of 3D printed concrete. Cem. Concr. Compos. **152**, 105657 (2024)
60. He, L., Li, H., Chow, W.T., Zeng, B., Qian, Y.: Increasing the interlayer strength of 3D printed concrete with tooth-like interface: an experimental and theoretical investigation. Mater. Des. **223**, 111117 (2022)
61. Van Der Putten, J., De Schutter, G., Van Tittelboom, K.: Surface modification as a technique to improve inter-layer bonding strength in 3D printed cementitious materials. RILEM Tech. Lett. **4**, 33–38 (2019)
62. Rasehorn, I.J., Mai, I., Stephan, D.: Impact of temperature, humidity and curing on printability, strength and shrinkage in 3D concrete printing. Humidity and Curing on Printability, Strength and Shrinkage in 3D Concrete Printing
63. Wei, Y., Zhang, H.: Influence of temperature and humidity on mechanical properties of calcined oyster-shell powder-modified 3D-printed concrete. ACS Omega **9**(37), 39180–39187 (2024)
64. Sun, B., Li, P., Wang, D., Ye, J., Liu, G., Zhao, W.: Evaluation of mechanical properties and anisotropy of 3D printed concrete at different temperatures. Structures **51**, 391–401 (2023)
65. Alchaar, A.S., Al-Tamimi, A.K.: Mechanical properties of 3D printed concrete in hot temperatures. Constr. Build. Mater. **266**(1), 120991 (2021)
66. Ma, L., Zhang, Q., Jia, Z., Liu, C., Deng, Z., Zhang, Y.: Effect of drying environment on mechanical properties, internal RH and pore structure of 3D printed concrete. Constr. Build. Mater. **315**, 125731 (2022)
67. Nienaber, L., Kanyenze, S.S., Combrinck, R. (ed.): Water Absorption Reduction of 3D Printed Concrete Using Silicones. Springer (2024)
68. Craveiro, F., Nazarian, S., Bartolo, H., Bartolo, P.J., Pinto, D.J.: An automated system for 3D printing functionally graded concrete-based materials. Addit. Manuf. **33**, 101146 (2020)
69. Vlachakis, C., Perry, M., Biondi, L., McAlorum, J.: 3D printed temperature-sensing repairs for concrete structures. Addit. Manuf. **34**, 101238 (2020)
70. Ozturk, O., Nair, S.D.: Mechanisms governing in-depth infiltration of crack filling solutions in concrete using a magnetic approach. **192**, 107856 (2025)
71. Shields, Y., Tsangouri, E., Riordan, C., de Nardi, C., Godinho, J.R.A., Ooms, T., Antonaci, P., Palmer, D., Al-Tabbaa, A., Jefferson, T., de Belie, N., Van Tittelboom, K.: Non-destructive evaluation of ductile-porous versus brittle 3D printed vascular networks in self-healing concrete. Cem. Concr. Compos. **145**, 105333 (2024)
72. Belie N., de Gruyaert, E., Al-Tabbaa, A., Antonaci, P., Baera, C., Bajare, D., Darquennes, A., Davies, R., Ferrara, L., Jefferson, T., Litina, C., Miljevic, B., Otlewska, A., Ranogajec, J., Roig-Flores, M., Paine, K., Lukowski, P., Serna, P., Tulliani, J.-M., Vucetic, S., Wang, J., Jonkers, H.M.: A review of self-healing concrete for damage management of structures. Adv. Mater. Int. **5**(17) (2018)
73. Lee, K.-J., Lee, J.-H., Jung, C.-Y., Choi, E.: Crack-closing performance of NiTi and NiTiNb fibers in cement mortar beams using shape memory effects. Compos. Struct. **202**, 710–718 (2018)
74. Zhang, Y., Zhu, X., Li, M., Zhang, C., Zhang, Y., Du, X., Banthia, N., Mechtcherine, V., Carstensen, J.V., Monteiro, P.J.M.: 3D printing technology in concrete construction. Nat. Rev. Clean Technol. 1–16 (2025)

Active Control of Cementitious Slabs and Roads

Eleni Korda, William P. Boshoff, Dengwu Jiao, Noor Yaseen, Dimitrios G. Aggelis, and Didier Snoeck

Abstract Materials capable of responses to external stimuli are reshaping infrastructure through active control strategies in concrete roads and slabs. This chapter examines how external triggers, such as water spraying, magnetic fields, electrical currents, or acoustic monitoring, can be employed to actively modify the properties of cementitious materials, thereby improving durability and functionality. Active control of plastic shrinkage cracking relies on capillary pressure sensors that activate rewetting interventions, significantly reducing crack formation. For self-desiccation and autogenous shrinkage, superabsorbent polymers (SAPs) provide internal curing, and their effectiveness can be extended through active control of desorption using acoustic emission monitoring. This enhances crack mitigation while minimizing negative impacts on mechanical strength. Magneto rheology offers adaptive vibration and stiffness control in slabs and pavements. Electrically conductive concrete, activated by external electrical stimuli, demonstrates promise for automated de-icing and

E. Korda · D. G. Aggelis
Department of Mechanics of Materials and Constructions, Vrije Universiteit Brussel, Ixelles, Belgium

E. Korda
Department of Structural Engineering and Building Materials, Ghent University, Ghent, Belgium

W. P. Boshoff
Faculty of Engineering, Built Environment and Information Technology, University of Pretoria, Pretoria, South Africa

D. Jiao
Department of Architecture and Civil Engineering, City University of Hong Kong, Hong Kong, China

N. Yaseen
Department of Civil Engineering, University of Engineering and Technology, Lahore, Pakistan

Irrigation Department, Peshawar, Khyber Pakhtunkhwa, Pakistan

D. Snoeck (✉)
Department of Building, Architecture and Town Planning, Université Libre de Bruxelles, City of Brussels, Belgium
e-mail: didier.snoeck@ulb.be

G. De Schutter and J. Sanjayan (eds.), *Active Control of Properties of Fresh and Hardening Cementitious Materials*, RILEM State-of-the-Art Reports 50,
https://doi.org/10.1007/978-3-032-30340-0_9

heating in cold climates, reducing reliance on salts and chemical agents. Finally, self-healing asphalt and cementitious systems utilize magnetic, ultrasonic, or induction-based triggers to initiate repair, supported by integrated sensing systems for targeted interventions. In conclusion, these strategies highlight a shift toward responsive infrastructure where materials can adapt and maintain performance. Future research must address challenges in standardization, scalability, and integration with digital monitoring and automation to fully enable self-maintaining, smart infrastructure systems.

Keywords Active control · Smart infrastructures · Shrinkage cracking · Magneto rheology · Conductive concrete · Healing concrete

1 Introduction

Smart materials that respond dynamically to external stimuli are reshaping the future of infrastructure. This chapter discusses the active control properties in slabs and roads. The 'active' character of the control involves the application of an external trigger signal that induces a response in the cementitious material, with a modification of properties or behaviour as a result. In the active control process, the intervention is based on observations and/or measurements (quantified), giving the necessary information to take a decision regarding required intervention. The first part of this chapter is related to Chap. "Active Control of Concrete Shrinkage and Associated Cracking" on shrinkage characteristics, and we would like to refer the reader to this chapter to have more background on shrinkage and shrinkage-related cracking phenomena. Relevant studies showcasing the principles and active control are showcased in the subsequent sections. This section explores the current and potential uses in adaptive slabs and road systems.

2 Active Control of Plastic Shrinkage and Related Cracking

Passive control of plastic shrinkage and hence reducing plastic shrinkage cracking by reducing evaporation has been around for many decades, at least since 1957 [1]. Active control of these types of cracking has first been published by Schmidt and Slowik [2] with some other publications following [3–6]. This active control is done by measuring the internal capillary pressure of fresh concrete and consequently reacting at the right times with wetting/spraying of the concrete with water. This has been shown to reduce [3] and prevent plastic shrinkage cracking [4].

This section explains the concept of active control of plastic shrinkage cracking, the sensors used and their limitations, and case studies of this active control approach being demonstrated.

2.1 Concept of Active Control of Plastic Shrinkage Cracking

Plastic shrinkage of fresh concrete is driven by capillary action due to the concrete drying because of evaporation, as explained in Sect. "Plastic Shrinkage and associated cracking" of Chap. 5. The reader is referred to Sect. "Plastic Shrinkage and associated cracking" for a more in-depth explanation of plastic shrinkage and associated cracking. This section focuses on the use of the sensors to monitor this plastic shrinkage cracking mechanism, namely internal capillary pressure.

There exists no knowledge or model to predict the actual capillary pressure values of concrete at a certain time. This highly depends on the mixture design, the environmental conditions, the settings regime and the age of the specific concrete. Therefore, a baseline will always have to be determined for the specific concrete. This could include different baselines at different evaporation rates as was done by Deysel et al. [4].

A baseline for a typical concrete is schematically shown in Fig. 1 for concrete that is uncured and cured [3]. It shows the results of the negative capillary pressure of two concretes deviating from each other at the point of curing onset, as to be expected with the steep increase of the capillary pressure values over time. It is assumed that the negative capillary pressure development is the ideal curve and the negative capillary pressure of the uncured concrete must be actively controlled to move towards the cured concrete's curve. This is done by the intervention of wetting/spraying the concrete with water when a certain negative capillary pressure threshold is reached. This process lowers the capillary pressure which increases again shortly after as evaporation continues. This process should continue until the desired time when the concrete has gained sufficient strength to resist the plastic shrinkage cracking. Typical threshold values used by the authors of Fig. 1 are −10 kPa [2] and −25 kPa [6]. No reference is, however, made to how the threshold value is to be determined.

Deysel et al. [4] determined the threshold by a simple assumption that the area below the negative capillary curve (the integral) of the concrete in the field must be equal to or less than the area that the area below the curve of the baseline that did not show plastic shrinkage cracking. This base line was determined by varying the evaporation rate over moulds that are inducing plastic shrinkage cracking in a controlled environment. The evaporation rate that does not result in cracking is used as the baseline and the associated negative capillary pressure curve is used to determine the threshold. Typical results of different evaporation rates and associated negative capillary pressure curves are shown in Fig. 2a while the cracking is shown in Fig. 2b. The drying time is when the bleeding water has evaporated and water is being evaporated from the concrete itself.

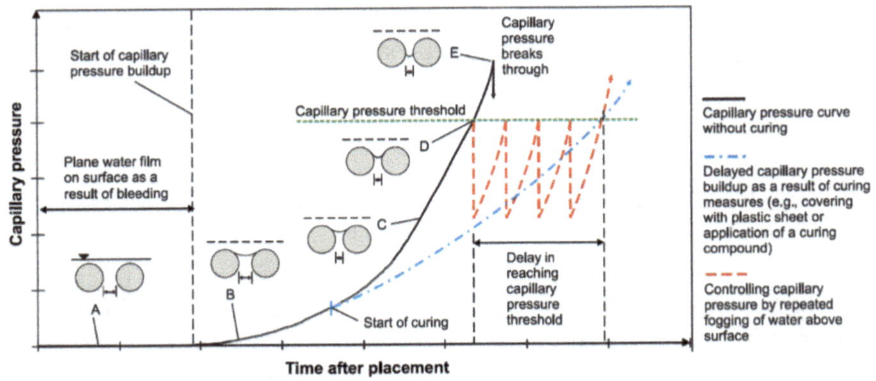

Fig. 1 Typical capillary pressure readings of concrete with and without curing [3]

Deysel et al. [4] proposed that the rewetting continues until after the initial set based on the proposition by Boshoff and Combrinck [7] that the water evaporated from the concrete up to initial set can be linked to the plastic shrinkage crack width at final set. To summarize, the concept is simple. The practitioner needs to keep the capillary pressure within a threshold by rewetting the concrete until the initial setting time or when the capillary pressure curve crosses the no-cracking capillary pressure curve.

2.2 Capillary Pressure Sensors

There are two types of commercial and non-commercial sensors available for the measurement of pore pressure values. The first type is an "open" type where the pressure sensor is directly exposed to the pore water system. They can typically not measure pressures greater than −100 kPa. This is because air can enter the system easily which is called the air entry value. The measurements by Schmidt and Slowik [2] typically showed air entry values of no more than −60 kPa. These sensors can normally be used with minimal preparation, i.e. at the most, filling them carefully with water. A design layout of these sensors can be seen in Fig. 3.

The second type is where a porous ceramic is used to shield the sensor from the capillary system which then becomes a medium and such a sensor is often called a tensiometer. They are especially useful in soil mechanics where the suction is measured in partially saturated soils. They can measure up, for example, to −2000 kPa [8]. These sensors are of course more complex to manufacture and require a strict preparation protocol, sometimes up to three days. They also need to be used immediately after they are removed from de-aired water. An example of a cross section of one of these sensors is shown in Fig. 4.

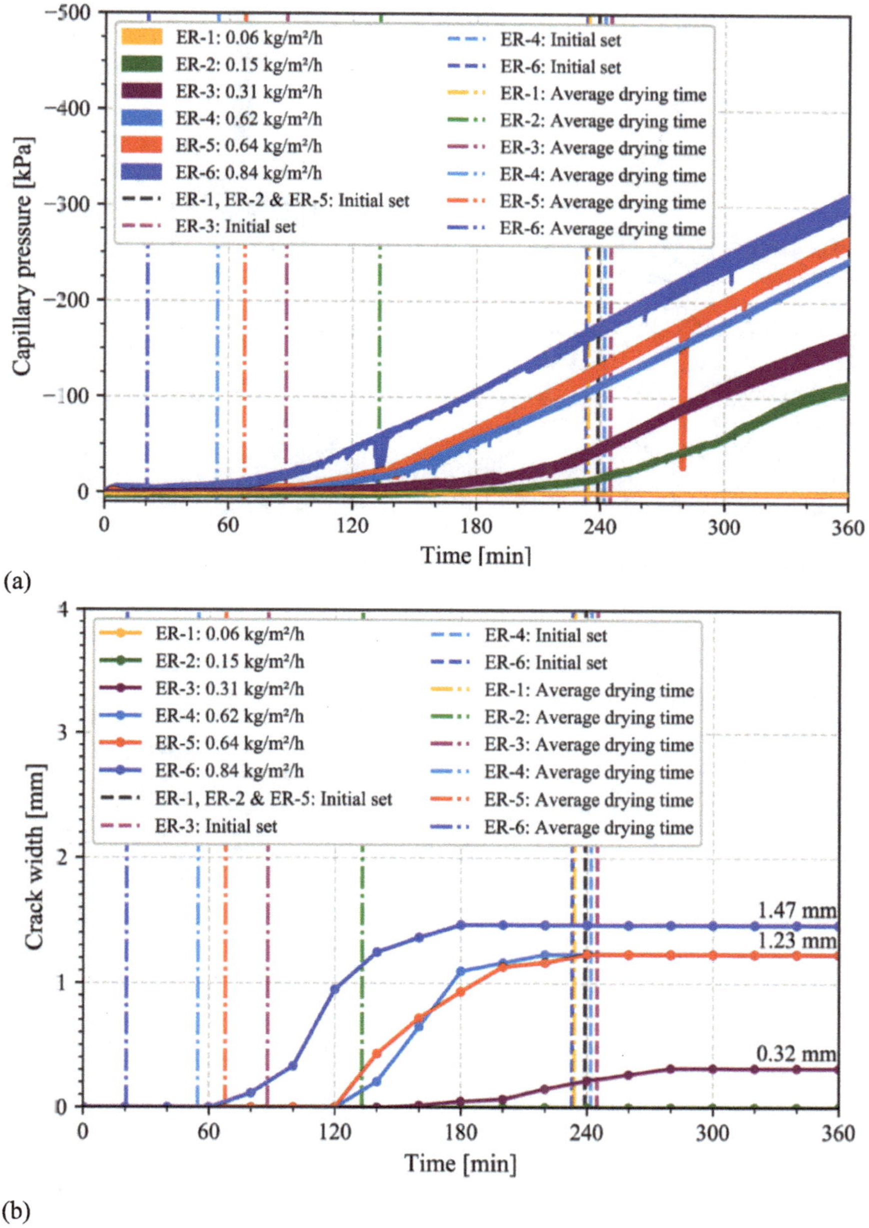

Fig. 2 **a** The capillary pressure development curves for different evaporation rates, and **b** The crack width developments at different evaporation rates [4]. Please refer to the original document for the figures in colored version

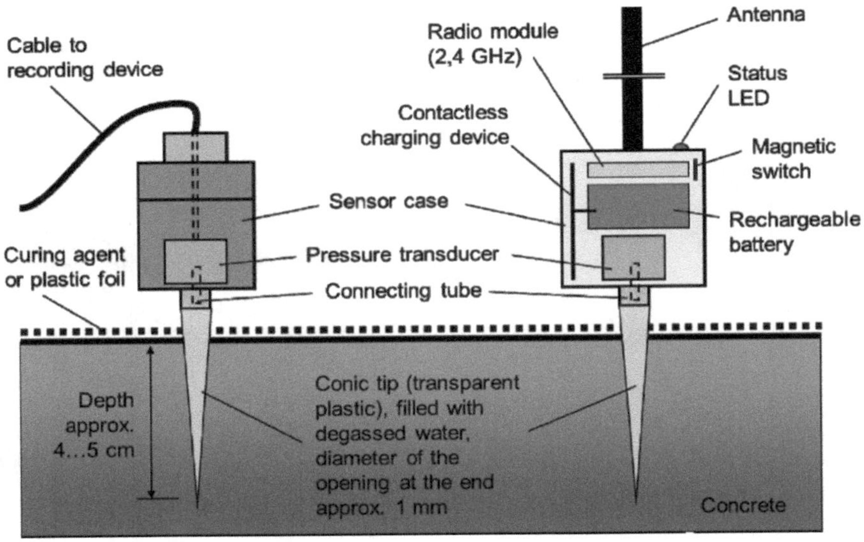

Fig. 3 Wired and wireless sensors for capillary pressure [6]

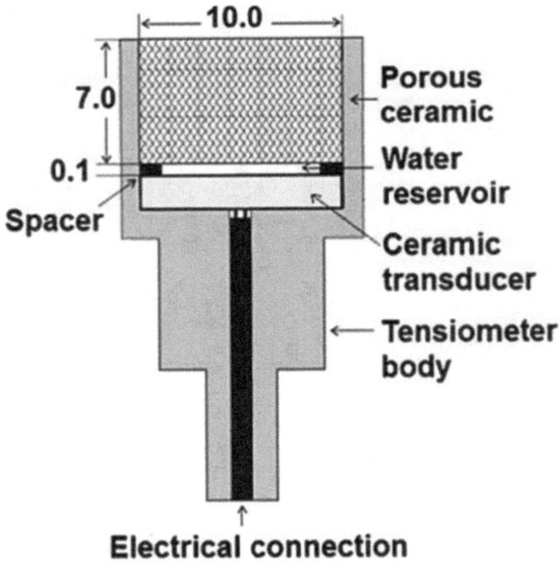

Fig. 4 A cross section of a high capacity tensiometer with a porous medium [9]

As the threshold for active control of the plastic shrinkage cracking is typically less than −50 kPa, the use of pore pressure sensors without a ceramic film will normally be sufficient for this application.

2.3 *Examples of Active Control of Concrete for Plastic Shrinkage*

Two examples could be found in literature of the use of active control of concrete for the prevention/reduction of plastic shrinkage cracking. The first example is the one by Schmidt and Slowik [2] where two slabs in the field were monitored for capillary pressure, but only one was rewetted when the pressure reached a threshold. In this case, the threshold was −10 kPa and the rewetting was done until 6.5 h after casting. The setup is shown in Fig. 5a, b. The graph showing the capillary pressure development both in the slab that was monitored and the one that was not monitored is shown in Fig. 5c. Finally, the measured crack lengths and width are shown in Fig. 5d, e for both the slabs, clearly showing the success in reducing the crack widths using this method.

As explained in Sect. 2.1, Deysel et al. [4] used an evaporation rate that shows no cracking as the benchmark. The authors then calculated the threshold for negative capillary pressure as a value to ensure the same area under the curve for capillary pressure. For the two demonstrations, they used moulds based on the ASTM C1579 standard for plastic shrinkage cracking. For the first demonstration, the benchmark evaporation rate was 0.15 $kg/m^2/h$ which resulted in a no-cracking capillary pressure of −5.08 kPa while the 0.52 $kg/m^2/h$ evaporation rate ended up in a threshold of −3.77 kPa. The lower threshold definition stems from the fact that higher evaporation rates accelerate capillary pressure build-up, thereby shifting the cracking benchmark to earlier times compared to lower evaporation rates. The resulting capillary pressure curves are shown in Fig. 6 for both Demonstration 1 and Demonstration 2. Both demonstrations showed no cracking after the active control intervention.

3 Active Control of Self-desiccation and Shrinkage Control

Self-desiccation is a result of chemical shrinkage taking place in the absence of additional water. As self-desiccation occurs, capillary pressure increases, and internal RH levels will decrease as cement hydrates. In low w/c mixtures, the drop in internal RH levels requires additional water to be provided to mitigate self-desiccation shrinkage. To address this, internal curing mechanisms such as LightWeight Aggregates (LWA) or SuperAbsorbent Polymers (SAPs) have been proven very effective [10]. The purpose of internal curing is to provide readily available additional water to keep the capillary porosity of the hydrating cement saturated, minimizing, therefore, autogenous stresses and strains [11]. In turn, the additional water may result in an increase of the degree of hydration and hence, increased strength and reduced transport coefficients. Conventionally, this additional water was provided through external curing techniques such as ponding, misting, fogging or covering of the concrete surface. The advancement of concrete technology led to the use of high-strength, low w/c ratio concretes, which are characterized by very low capillary porosity, preventing

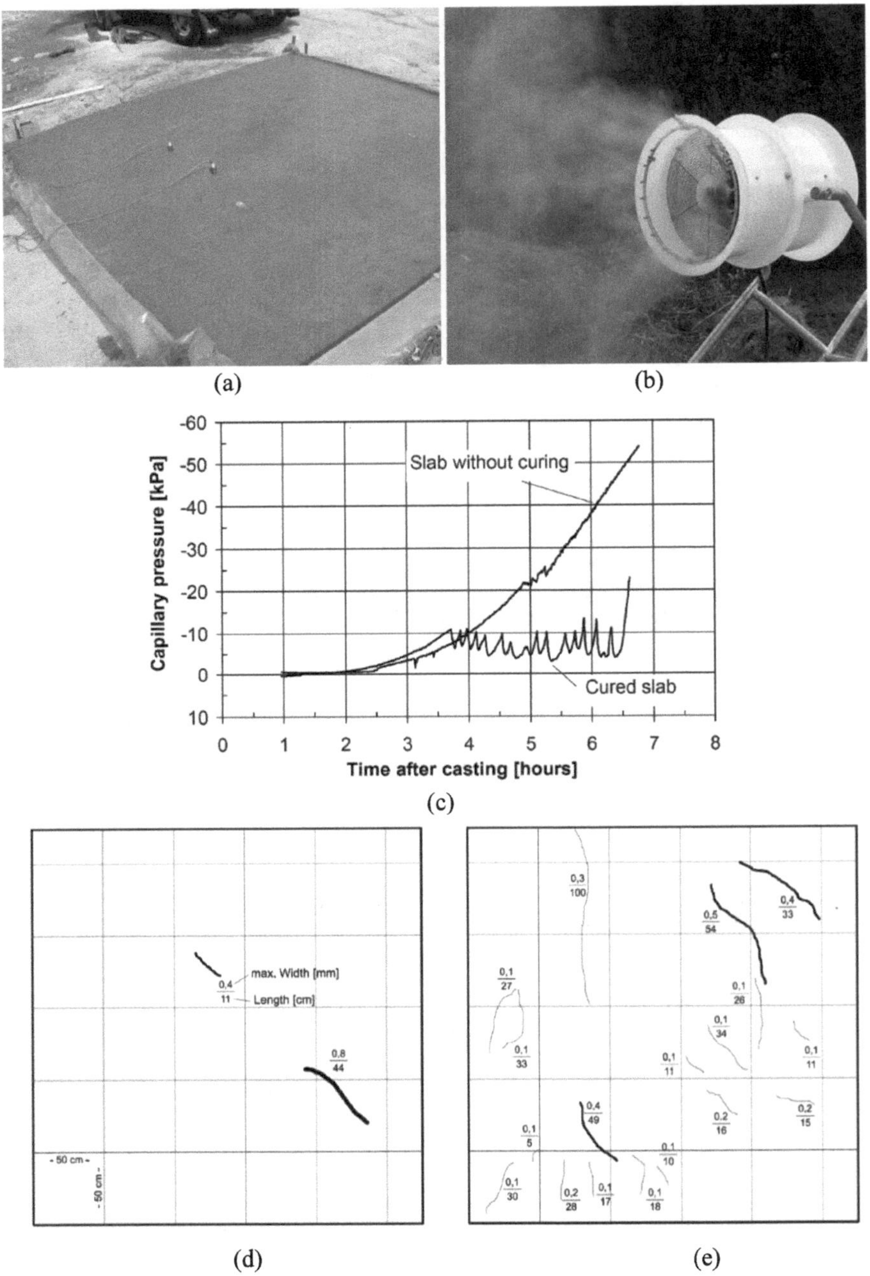

Fig. 5 **a** The positioning of the sensors in the slab, **b** the fog spray used for rewetting, **c** capillary pressure development for the slab with and without rewetting (curing), and crack widths of the slab **d** with and **e** without rewetting (curing) [2]

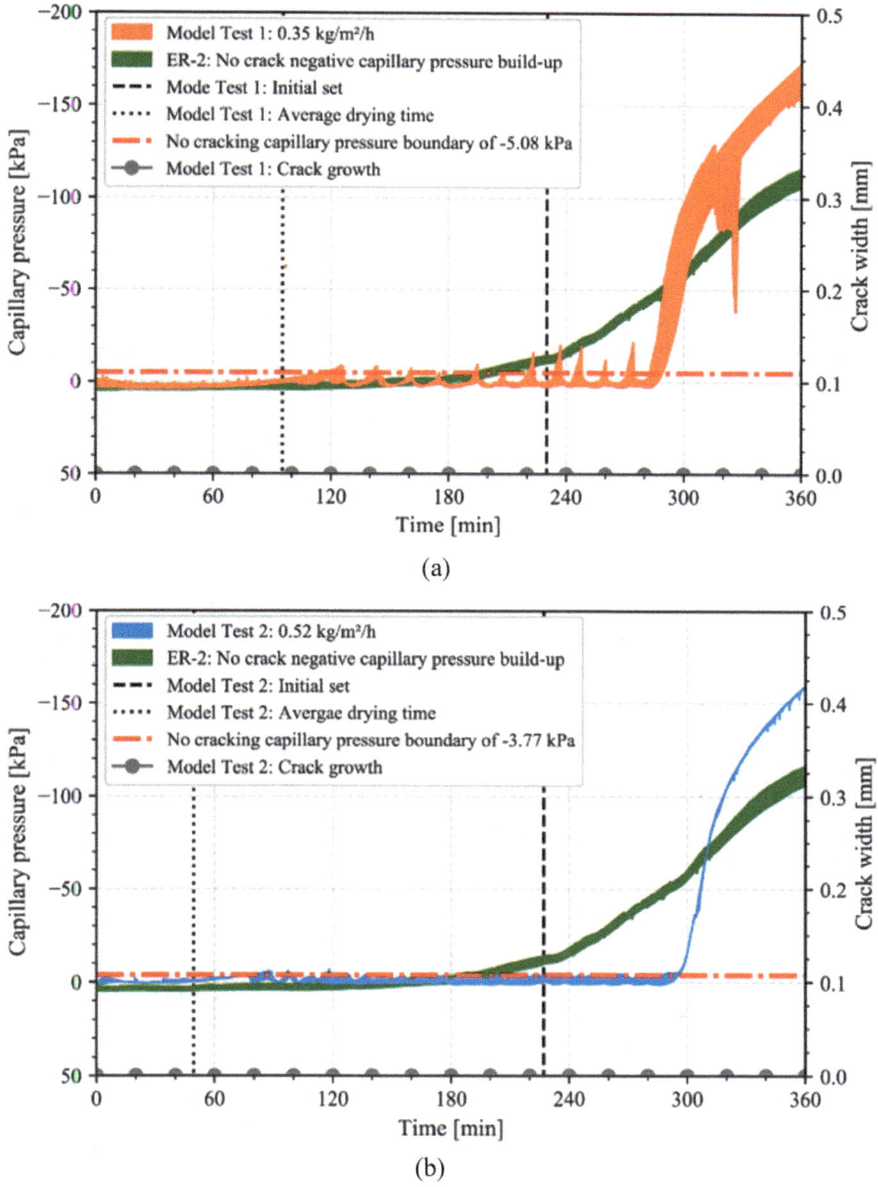

Fig. 6 The capillary pressure response curves for two demonstrations **a** and **b** with the evaporation rates shown in the first entry of the legend [4]. Please refer to the original document for the figures in colored version

external curing water from penetrating deep into the matrix. Therefore, the surface remains saturated, while the bulk volume undergoes significant self-desiccation [11].

3.1 Principle of Internal Curing and Superabsorbent Polymers

Internal curing can provide the necessary water quantity throughout the self-desiccation process, with the proper rate (when exposed to ambient environmental conditions) and a proper homogeneity within the bulk volume, resulting in a well-saturated and stress-free cementitious matrix. In this manner, early-age cracking is minimized. SAPs, for example, are added in concrete in a dry state and permit free design of the pore shape and size during concrete hardening [12]. For the aforementioned reasons, SAPs have been widely used for mitigating autogenous strains and self-desiccation [13–17]. SAPs are added to concrete to change the rheology [18], mitigate plastic [19] and autogenous shrinkage [12], improve the freeze-thawing resistance [20], improve sealing [21] and healing [22] characteristics. However, by including them in the mixture design, after desorption and internal curing, they will leave behind macro pores, hence decreasing the overall mechanical properties [15] especially in mixtures with high water-to-binder ratios and in high amounts [10, 23].

An interlaboratory study of 13 institutes investigated the autogenous shrinkage evolution of HPC (w/c of 0.3 and 10% per cement weight of silica fume inclusion) enhanced with 0.3 m% SAPs, showing an autogenous shrinkage mitigation of 109%, 84% and 73% during the first three days, respectively, and 58%, 52% and 42% at later ages of 7, 14 and 28 days, respectively [24]. The timely delivery of water by the SAPs, although beneficial, is highly dependent on the SAP type and size, as well as the environmental conditions to which concrete is exposed [25]. SAPs can be categorized as either self-releasing or retentive [19, 26, 27]. The former releases the entrained water without any external trigger after casting, or due to vibration or compaction and can lead to an increase in the effective w/c ratio, increasing plastic shrinkage and the associated cracking risk. The latter retains the water until it can successfully be used for internal curing during the self-desiccation process, mitigating, therefore, autogenous shrinkage. The releasing behaviour is based on the anionic functionalities present in an SAP's internal polymer network, which interact with cations present in the cementitious pore fluids. These interactions, particularly with K^+, Na^+, and Ca^{2+}, accelerate fluid desorption and lower their absorption capacity. Even though all SAPs can mitigate autogenous shrinkage to some extent, retentive SAPs continue to mitigate shrinkage for longer times (up to 7 days) compared to self-releasing SAPs [26].

Another important factor influencing SAP desorption is the environmental conditions. As hydration progresses and ions accumulate in the matrix's pore solution, the SAPs release entrained water back into the matrix [28]. This occurs through increased surface tension and capillary pressure, as well as alterations in the ion concentration kinetics, which directly impact the SAP desorption rate. In unsealed systems, evaporation leads to premature depletion of free water, causing SAPs to release their stored water more quickly than in sealed systems. Elevated temperatures and windy conditions have the same effect, increasing the risk of early-age shrinkage and resulting cracking. For these reasons, it is essential to detect and control SAP desorption to ensure optimal internal curing performance. Controlled SAP desorption can optimize curing and help maintain SAP pores saturated for a longer period, reducing thermal deformation and therefore, cracking [29]. It is not only beneficial for preventing early internal curing water exhaustion but also for optimizing curing without compromising mechanical properties. The amount of added SAP directly impacts compressive strength through increased (macro)porosity. Lower dosages of SAP will have a less detrimental effect on compressive strength; however, they will be less effective in mitigating shrinkage.

3.2 *Using Active Control for Curing Processes*

The long-term performance of concrete relies heavily on the early-age curing processes, yet current methods provide limited opportunities to adjust its properties after mixing/casting. Active Control (AC) of concrete properties addresses this limitation by enabling specific concrete property adjustments during the curing and hardening stages [30]. Crack control is particularly important for High-Performance Concrete (HPC), which is prone to (early-age) shrinkage due to its low porosity [31]. As the demand for HPC structures grows, effective early curing control becomes even more critical. In practice, one extra dollar that can be used for controlling the curing of the material (e.g. to prevent cracking) would save 25 dollars when extensive repair would be necessary after a certain operational life [32]. Cracks originating from early-age shrinkages, such as thermal, plastic, or autogenous, are more randomly distributed and, therefore, more crucial to concrete's long-term durability than mechanically induced cracking, which is normally discrete, concentrated, and hence easier to detect and repair [33]. SAPs can mitigate shrinkage; however, their effectiveness depends on the rate of cement hydration and the environmental conditions. Premature water desorption or excessive drying can impair SAP behaviour, increasing the need for (pro)active curing control methods to ensure desirable SAP performance. Therefore, a methodology to actively control the desorption behaviour of SAP would enhance their performance by extending internal curing and offering prolonged protection during the sensitive early stages of material development, even allowing for a reduced SAP dosage. This is especially important for large surface area elements, such as slabs and roads, which are particularly prone to rapid drying and early-age cracking. The AC methodology presented here, known as hygral intervention, combines external water curing with internal curing water provided by SAPs, creating a synergistic curing mechanism. External water acts as an active intervention, while SAPs passively provide internal curing by responding to capillary and

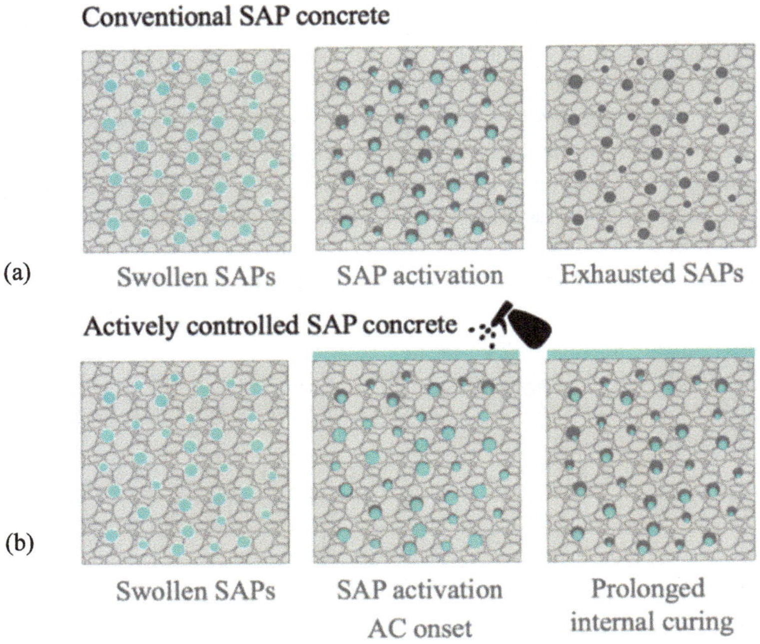

Fig. 7 Schematic representation of AC effect on internal curing. **a** Conventional internal curing and **b** internal curing combined with AC. Redrafted with permission from [34]

osmotic pressures. More specifically, the AC timeframe can be thought of as a period of controlled RH levels, externally (wet surface), but also internally since the free matrix water is prevented from evaporating, causing SAPs to maintain their entrapped water during the controlled period, as shown in Fig. 7.

Although active control of SAP desorption provides an effective and sustainable way to improve early-age concrete properties and durability, it requires accurate monitoring to identify the right moments for the hygral intervention. Acoustic Emission (AE) monitoring is pivotal in this sense, with its high sensitivity, reaching the attoJoule (10^{-18} J) level, enabling the detection of microstructural changes [35]. Significant improvements have been recently achieved in monitoring fresh and hardening cementitious materials using AE [35–37]. AE is particularly useful for collecting real-time data from the interior of the material and providing information as soon as a few minutes after casting. Parameters such as the cumulative acoustic activity (number of recorded signals or "hits"), absolute energy, amplitude, and frequency content of the waveforms can be used to detect and interpret different curing stages and cracking modes [38, 39]. A summary of the research conducted on monitoring fresh concrete with AE is presented in [35], while a typical AE setup for fresh concrete monitoring can be seen in Fig. 8. In this application, piezoelectric sensors are attached to the external side of the mould recording the transient elastic waves that reach their surface. These are pre-amplified and transformed into digital electric waveforms in the acquisition board.

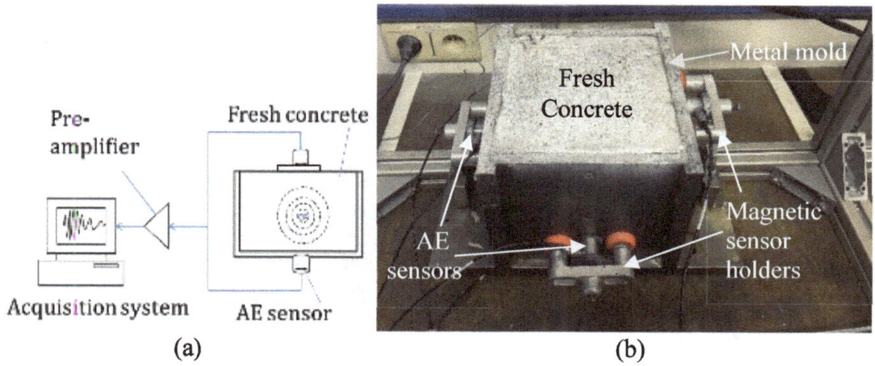

Fig. 8 **a** Schematic representation of an AE setup for fresh concrete monitoring, **b** photograph of an indicative setup. Redrafted with permission from [34]

AE sensors can monitor any mechanisms that are related to irreversible changes or displacement like settlement, aggregate movement, bubble release, shrinkage cracking and hydration [37, 40, 41]. Among the aforementioned mechanisms, AE can reliably detect SAP desorption onset after final setting by monitoring the acoustic cumulative activity, which shows a considerable increase in the case of SAP concrete indicating water release [39, 42], as can be seen in Fig. 9a. As has been investigated in Korda et al. [43, 44] using SEM, the AE activity is connected to the shrinkage of the SAP that causes detachment of the SAP particles from the internal surface of their cavity. Since the onset of the SAP activity becomes evident (between 13 and 14 h in Fig. 9b), at this point, thin layers of water are sprayed onto the specimen, deactivating desorption. As shown in Fig. 9b, after the moment of external water application, the AE curve flattens demonstrating the immediate pause in the SAP action and the direct way of controlling the SAP behaviour. When SAP particles start to activate again (at approximately 16 h in Fig. 9b) another spraying takes place, while this external intervention can take place repeatedly. This deactivation/reactivation method has the possibility to extend internal curing, further reduce shrinkage cracking and simultaneously allow to use lower SAP dosages without compromising concrete strength [44]. The external curing offered by the surface water layer and the internal curing by the SAPs create a synergetic mechanism of internal and external curing to maintain internal RH levels by preventing evaporation and maintaining the capillary pores saturated. Figure 9c shows typical AE curves from various concrete compositions, where the extension of the SAP activity for several hours is demonstrated in the case of AC of the SAP desorption, offering protection against early age cracking for longer period.

By adopting the AC method using AE, the near-the-surface shrinkage cracking is mitigated for SAP concrete, as shown in Fig. 10. The SEM pictures concern samples taken 10 mm below the surface of the specimens. The use of SAP proved effective in decreasing the total microcracking fraction area to almost half compared to reference concrete (REF). Still, the actively controlled specimen (SAP_W) exhibited a

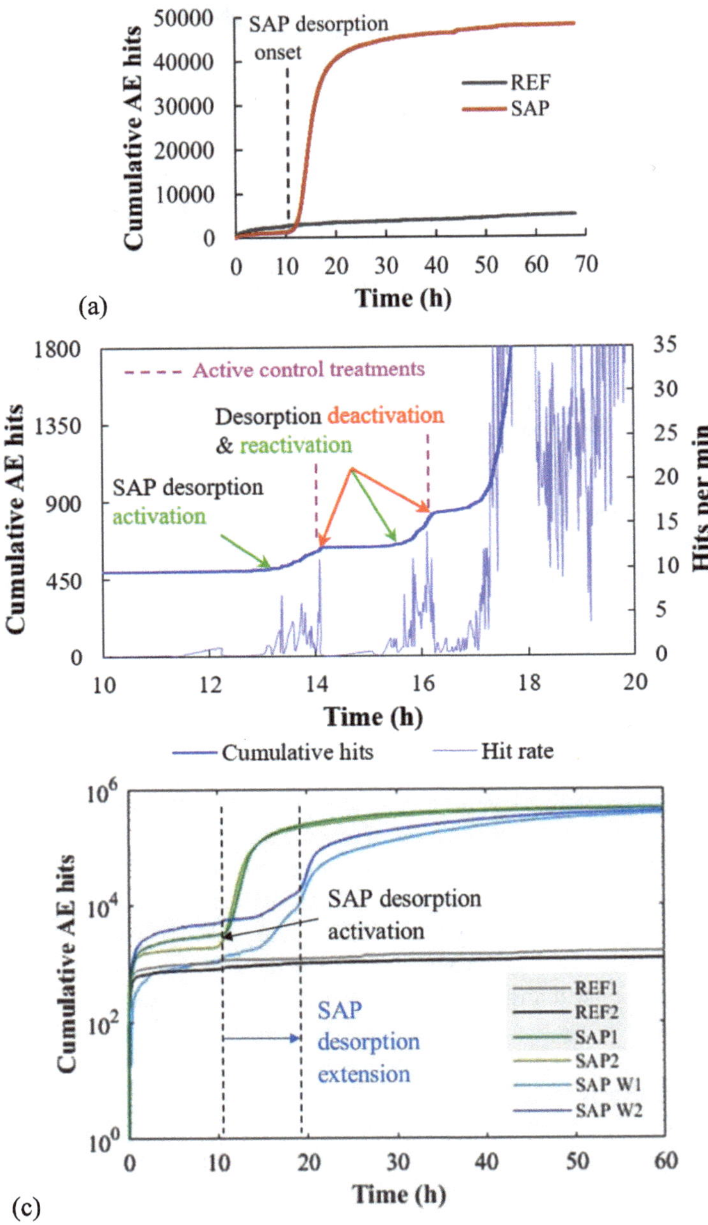

Fig. 9 **a** AE activity of conventional concrete (REF) versus concrete with SAPs (SAP), **b** Active control of SAP desorption rate: cumulative AE activity and hit rate. The green arrows represent the SAP desorption onset and reactivation. The red arrows represent the deactivation of SAP desorption, and **c** Cumulative AE activity of REF, SAP and SAP_W specimens (numbers 1 and 2 represent two specimens). REF refers to the reference composition, SAP refers to a composition containing 0.2 m% per cement weight of SAPs and SAP_W refers to the actively controlled SAP specimen [44]

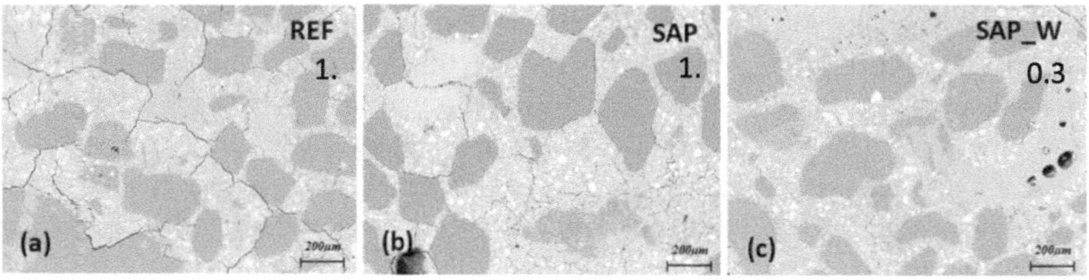

Fig. 10 Near-the-surface microcracking fraction area: **a** conventional concrete, **b** SAP concrete, and **c** SAP_W concrete [44]. The scale bar amounts to 200 μm

much stronger decrease in cracking area to about 1/5 compared to the REF (0.33% compared to 1.72%) [44].

The AC methodology takes advantage of the unique ability of SAPs to swell and shrink repeatedly based on surrounding water availability. Importantly, AC optimizes SAP concrete curing without requiring high SAP content for shrinkage mitigation, which would compromise compressive strength. Finally, different amounts of externally applied water will have a different impact on the SAP desorption rate, CP, the AE activity, and the microstructure. The proposed methodology has great potential to be upscaled to slabs and roads and improve the durability performance of concrete. One point of attention is the inherent damping of the fresh viscous concrete. For an event occurring towards the centre of the specimen (or far away from the sensors), the attenuation of approximately 2 dB/mm may not allow recording depending also on the original amplitude [45]. AE events occurring closer to the mould are more likely to be recorded as the accumulated attenuation becomes significantly lower and the steel mould, exhibiting much less damping (less than 0.1 dB/mm), acts as a waveguide transmitting signals from a larger volume to the sensor. In any case, the high damping necessitates the use of very sensitive transducers. In addition, industrial formworks do not always consist of metal, but also of wood or plastic. This would make AE monitoring of the early-age concrete more challenging, due to increased damping, but, depending on the formwork, not impossible as the increased SAP activity (even with lower amplitude) could still be identified and point out the times of the active intervention.

4 Magnetorheological Response in Slabs or Roads

Magneto-Rheological materials, including Fluids (MRFs) and Elastomers (MREs) [46, 47], are smart composites whose rheological properties, such as viscosity and stiffness, can change reversibly under external magnetic fields. MRFs are composed of micron-sized ferrous particles suspended in a non-magnetic carrier fluid, whereas MREs embed these particles within an elastic matrix. Upon exposure to an external magnetic field, the ferrous particles will be polarized to align into chain-like structures along the field direction, leading to the material transforming from liquid-like to a semi-solid state [48, 49]. As a result, the yield stress can rise to 100 kPa and the viscosity could increase magnitudes of 105–106 times of its original value [50, 51]. Details about the excitation mechanisms can be found in Chap. "Active Control of Rheology of Cementitious Materials". The rapid response (typically milliseconds) and tuneable mechanical behaviour of MR materials make them highly promising for dynamic applications in civil engineering [52]. This section mainly discusses their possible applications in adaptive infrastructure like slabs and roads.

MR materials show immense potential in mitigating vibration issues in slabs induced by dynamic forces such as train movements or seismic activity. Koo [53] pioneered the use of MR dampers in semi-active tuned vibration absorbers to control structural vibrations in scenarios like building floors. Further simulations by Li et al. [54] using MATLAB found that connecting MR isolators in parallel in long beams effectively countered excitation forces from moving loads. This setup reduced vibration forces transmitted to the ground from the slab track by 31.44%, beneficial to achieving the on-line adaptive optimization for the parameters of stiffness and damping. Wei et al. [55] advanced the concept with a semi-active MR damper-steel spring floating slab track system. The findings revealed that the introduction of MR damper would not compromise the subway vehicle safety while achieving enhanced vibration reduction, evidenced by a roughly 10% decrease in the vertical maximum dynamic displacement. Moreover, da Silva Brandão et al. [56] validated semi-active MR dampers as powerful tools for vibration control in tall buildings, particularly under wind-induced dynamic loads.

To improve driving conditions and address road degradation issues, Arsenault [50] examined the possibility of incorporating MR fluid dampening components to prepare rigid pavement joints. While full validation of the concept is pending, initial results indicated enhanced stiffness and effectiveness of joints under dynamic vehicular loads. Similarly, MREs show promise for adaptive traction control on icy pavements. Specifically, pavement surfaces embedded with MREs could dynamically adjust roughness in response to magnetic fields, thereby improving safety and functionality in adverse conditions [51].

The transformative potential of MR materials for adaptive slabs and roads is undeniable. However, challenges such as optimizing material and system design, ensuring stability, achieving desired responsiveness, and maintaining long-term durability, are main problems that must be overcome in real construction applications. To fully realize their potential of MR-based smart roads and slabs, future research should focus on extensive field trials, eco-friendly formulations, and cost-effective implementation strategies. In a word, MR-enabled smart infrastructure holds the promise of a safer and more adaptable future for civil engineering.

5 Electrically Conductive Concrete for De-Icing and Heating

Electrically Conductive Concrete (ECC) has emerged as a promising solution for de-icing and heating applications, transforming conventional construction material into a multifunctional infrastructure component. This approach allows for the modification of concrete properties and performance in response to quantified observations and measurements. While the broader potential of ECCs extends to domains such as structural health monitoring, electromagnetic shielding, and smart sensing, the scope of the present discussion is specifically confined to de-icing and surface heating applications in concrete roads and slabs.

To enable active control of electrically conductive concrete for applications in slabs and roads, researchers have investigated the use of external electrical trigger signals. The primary mechanism underpinning this functionality is Joule heating, in which electrical resistance within the conductive matrix generates heat when an electric current is applied. This property is particularly critical for infrastructure exposed to cold climates, where surfaces such as bridge decks, pavements, and airport runways benefit from automated, energy-efficient snow and ice removal without relying on chemical de-icers or mechanical scraping. This active control strategy involves the application of an external stimulus to induce a targeted response in the cementitious material.

Building upon this principle of externally triggered heating, several studies have explored how such active control strategies can be systematically optimized and scaled for diverse infrastructure applications, especially where conductive fillers are involved. A particularly promising method involves self-heating ECC systems, in which case conductive fillers, including carbon fibres [57–60] and carbon nanotubes [61, 62], are dispersed within the cementitious matrix. When an external power source is connected, electric current passes through this conductive network, generating uniform heat. This self-heating behaviour can be precisely controlled using variable current or voltage, enabling responsive and localized de-icing even during active snowfalls [63].

Various studies have demonstrated the feasibility of actively controlling concrete properties by applying an external electrical stimulus to electrically conductive concrete, with or without conductive fillers, in applications such as roads, airport runways and slabs. For example, Li et al. [64] proposed a compelling self-de-icing road system integrating a Carbon NanoFiber Polymer (CNFP) thermal source with a MultiWall Carbon NanoTube (MWCNT)/cement-based thermal conduction layer. The CNFP, with a stable resistivity of 0.72 Ω mm, demonstrated rapid heating and could generate a consistent heat flux of 1000 W/m^2 under an applied current of 2.3 A. The 3% MWCNT/cement composite exhibited significantly enhanced thermal conductivity (2.83 W/m K) compared to conventional cement-based materials. De-icing efficiency was influenced by variables such as ice thickness, ambient temperature, and heat flux, with optimal heating conditions identified using a time-energy index (β). Experimental evaluations confirmed the system's high de-icing and snow-melting efficiency, low energy consumption, and strong potential for application in roads, highways, and airport infrastructure. While the study makes a strong case for scalability and real-world viability, further investigation into long-term durability and cost under field conditions would strengthen its promise for widespread adoption.

Complementing nanocomposite-based approaches, Lai et al. [65] investigated a stronger structural integration method using carbon fibre grille systems, aimed specifically at large-scale pavement applications such as airport runways. Their field experiments, conducted in Beijing under ambient temperatures of −3 to −1 °C, demonstrated that an input power of 350 W/m^2 raises pavement surface temperatures by 4.63 °C, melting a 2.7 cm snow layer within two hours. Notably, residual heat within the pavement prolonged snow-melting efficacy even after power was cut, enhancing energy efficiency. The system's configuration—grilles embedded 5 cm below the surface with 10 cm spacing—proved effective, though the authors caution that induced thermal gradients may exacerbate structural issues like cracking or reinforcement corrosion. Although the feasibility is clear, their study calls for further research on long-term structural impacts, electromagnetic interference with aircraft systems, and practical implementation at scale.

Zhao et al. [66] validated the practicality of embedding Carbon Fiber Heating Wire (CFHW) in full-scale bridge-deck slabs under natural winter conditions. CFHW were longitudinally wrapped around the reinforcing mesh with an interval of 100 mm and positioned 40 mm deep within a 250 mm thick slab. Temperature data within the slab was collected using resistance sensors embedded at 50 mm depth intervals below the surface. The CFHW ends were connected to electric wires, and the power input to the slab was regulated using an AC booster. Field trials were run both for heating and de-icing experiments. The heating experiments demonstrated the average temperature increase on the slab surface over 6 h under varying input powers, showing a direct correlation between power input and temperature rise. Even upon power shutdown, the slab's internal temperature continued to rise briefly, revealing that the maximum temperature occurred at a depth of 100 mm. The de-icing experiments demonstrated that slabs with CFHW could melt 10 mm, 40 mm, and 70 mm snow layers in approximately 0.75 h, 2.5 h, and 3.5 h, respectively, when powered by the power input (500–800 W/m^2). These field trials, conducted at ambient temperatures down to −5 °C, confirmed that CFHW systems maintain stable electrical resistance and generate sufficient heat to raise surface temperatures above freezing within 1 h. This technology demonstrated direct field applicability, highlighting CFHW's reliability, energy efficiency, and durability as a viable solution for bridge-deck de-icing, while avoiding the environmental drawbacks associated with salt or complex fluid systems.

To actively control the heating process, researchers have explored the use of temperature sensors as triggers [67, 68]. In field application research by Tuan and Yehia [68], a microprocessor-based controller was integrated with a temperature-sensing unit to monitor conductive concrete slabs used for bridge deck de-icing. The system was designed to automatically activate power when slab temperatures dropped below 4.5 °C, and deactivate power when temperatures exceeded 12.8 °C, ensuring effective and energy-efficient de-icing. This automated control, driven by real-time thermocouple feedback, highlights the practical application of temperature sensors as active control inputs for managing electrically conductive concrete systems in cold climates. These sensors can be linked to real-time weather and traffic data, enabling the concrete to respond dynamically to changing environmental conditions and infrastructure needs. Such types of integrated smart infrastructure approaches allow for optimized performance and efficient energy usage, further enhancing the viability of electrically conductive concrete as a multifunctional construction material.

While the principle of active control of concrete properties by electrical signals has been put forth [67, 69–71], the underlying mechanisms explaining the observed effects of electrical stimulation on hydration and microstructure development remain not fully understood. This knowledge gap hinders the design and optimization of active control strategies for ECCs. Research is needed to investigate the fundamental mechanisms that govern the interaction between electric fields and cementitious materials. One critical factor influencing the performance of electrically conductive concrete is the effect of the applied electrical stimulus, such as voltage or current, on cement hydration kinetics, thermal expansion (leading to potential cracking), and the electrical conductivity of the embedded conductive materials.

Ultimately, the development of electrically conductive concrete is interesting for a wide range of applications, from de-icing of roads and bridges to the heating of building envelopes and industrial processes. Further advancements in material design, manufacturing techniques, and integration with smart infrastructure systems are expected to drive the widespread adoption of this innovative construction technology.

6 Self-healing Concrete/Asphalt Using Triggered Response

The study of the self-x properties of building materials is a recent research topic [72, 73]. One example is asphalt, where induction or microwave heating can be used to cure or heal observed cracking phenomena [74, 75], also in combination with compressive loading [76]. To this goal, several additives are added to enhance this effect. The healing is dependent on various factors [74, 75, 77–79], including the mixture design, bitumen viscosity, environmental factors (temperature), as well as the repair repetition, rest period, age, load characteristics, crack size and severity. As far as active control is concerned, magnetite powder [80] or iron-based particles that react to a magnetic impulse are used as an additive to heat road pavements and thus close cracks [81], using the hyperthermia effect [82]. There is an optimal size of particles as larger particles (more than 1 mm) show inhomogeneous heating and high temperature gradients during an alternating magnetic field. Other examples include ferrite [83, 84] or steel fibres [85, 86]. These principles are explained in detail in Chap. "Active Control of Rheology of Cementitious Materials".

Structural health monitoring systems can be used as an initial trigger for healing action [87]. These sensors detect cracks through acoustic emission or changes in resistivity, linked to a healing action. The healing effect can be combined with a self-sensing ability using various techniques such as acoustic emission [88–90], or more visual approaches. Ultrasonic waves can also be used as a trigger, for example in self-healing cementitious materials containing microcapsules [91]. In the latter, the size and overall distribution of the microcapsules, as well as the frequency, power, and time of ultrasonic treatment, play a role. Another indication could be the appearance of corrosion, by means of nickel-based multi-reference electrodes [14], to show when immediate repairs are needed. This data can feed into a control system that activates healing processes precisely where needed, improving efficiency and reducing material waste. Preventive measures are taken in terms of repairing asphalt roads by means of heating actions, and inserted sensors can indicate when this is necessary.

7 Conclusions and Future Prospects

This Chapter has discussed the current active research on active control of properties in roads and slabs. It is shown that the plastic shrinkage cracking can be controlled by interventions. This was done by rewetting the concrete when the capillary pressure reached a certain negative threshold. Different approaches have been used in the past, with success. One could argue that all concrete must be rewetted constantly with a fog spray until well past initial set. However, research showed that rewetting could be done in a more economical manner or is not even necessary in certain conditions for the purpose to prevent plastic shrinkage cracking.

However, several limitations and open questions remain. There is currently no standardized method for determining the capillary pressure threshold, and existing methods rely on empirical or semi-empirical assumptions such as area under the curve matching. Moreover, while commercial sensors have proven adequate for pressure ranges relevant to plastic shrinkage, their long-term durability, calibration needs, and integration into large-scale construction projects need further investigation.

AC can also be applied for internal curing purposes where external curing is activated once SAP particles start to de-swell and release their stored entrained water, as monitored by means of AE. It was shown that the related shrinkage cracking was reduced. The integration of internal curing with actively controlled SAP desorption may enhance early-age concrete durability. By extending internal curing through external hygral interventions and using real-time monitoring via AE, the AC methodology allows for reduced SAP dosages without compromising mechanical properties or durability. The synergistic mechanism of external and internal curing not only mitigates early-age shrinkage and microcracking but also paves the way for more sustainable and cost-efficient concrete roads and slabs. The technique is recently being studied, and further development is on the way.

Magnetorheological materials bring exciting prospects to the field of adaptive infrastructure. Their ability to respond to environmental stimuli in real-time enables dynamic control over vibration, stability, and traction. While promising results have been demonstrated in simulation and laboratory testing, real-world deployment remains in early stages. MR-enabled infrastructure represents a leap toward responsive civil engineering solutions.

Furthermore, the future with further integration of artificial intelligence and automation may enhance the active control capabilities of concrete roads and slabs. This may even allow for self-maintenance and improve the long-term performance and sustainability. The development of smart materials such as electrically conductive concrete and self-healing building materials represents a significant step toward sustainable infrastructure. Active control strategies, such as temperature-triggered systems and integrated sensors, enhance electrically conductive concrete responsiveness and enable smart, automated surface heating with minimal energy waste. By using external stimuli and integrating sensors for structural health monitoring, these materials provide proactive solutions. Their ability to respond autonomously or through controlled activation not only improves longevity and safety but also aligns with broader goals of reducing resource consumption and environmental impact for application in roads and slabs.

References

1. Lerch, W.: Plastic shrinkage. ACI J. **53**, 797–802 (1957)
2. Schmidt, M., Slowik, V.: Capillary shrinkage cracking and its prevention by controlled concrete curing. In: Concrete Durability and Service Life Planning—ConcreteLife'09, Haifa, pp. 23–30 (2009)
3. Slowik, V., Schmidt, M., Kässler, D., Eiserbeck, M.: Capillary pressure monitoring in plastic concrete for controlling early-age shrinkage cracking. J. Transp. Res. Board **2441**(1) (2014)
4. Deysel, R.C., Boshoff, W.P., Smit, M.S.: Implementing capillary pressure control measures to prevent plastic shrinkage cracking in concrete. Constr. Build. Mater. **397**, 132407 (2023)
5. Slowik, V., Hübner, T., Schmidt, M., Villmann, B.: Simulation of capillary shrinkage cracking in cement-like materials. Cement Concr. Compos. **31**(7), 461–469 (2009)
6. Schmidt, M., Slowik, V.: Capillary pressure controlled concrete curing in pavement construction. In: Airfield and Highway Pavement 2013: sustainable and Efficient Pavements, pp. 295–306. ASCE, Los Angeles (2013)
7. Boshoff, W.P., Combrinck, R.: Modelling the severity of plastic shrinkage cracking in concrete. Cem. Concr. Res. **48**, 34–39 (2013)
8. Jamali, A., Mendes, J., Nagaratnam, B., Lim, M.: A new four stage model of capillary pressure in early age concrete: insights from high capacity tensiometers. Cem. Concr. Res. **161**, 106955 (2022)
9. Toll, D.G., Lourenço, S.D.N., Mendes, J.: Advances in suction measurements using high suction tensiometers. Eng. Geol. **165**, 29–37 (2013)
10. Schröfl, C., Erk, K.A., Siriwatwechakul, W., Wyrzykowski, M., Snoeck, D.: Recent progress in superabsorbent polymers for concrete. Cem. Concr. Res. **151**, 106648 (2022)
11. Bentz, D.P., Weiss, W.J.: Internal curing: a 2010 state-of-the-art review (2011)
12. Jensen, O.M., Hansen, P.F.: Water-entrained cement-based materials I. Principles and theoretical background. Cem. Concr. Res. **31**(4) (2001)
13. Assmann, A., Reinhardt, H.W.: Tensile creep and shrinkage of SAP modified concrete. Cem. Concr. Res. **58**, 179–185 (2014)
14. Tenório Filho, J.R., De Belie, N., Snoeck, D.: The effects of internal curing and shrinkage cracking avoidance on the corrosion of reinforced concrete walls with superabsorbent polymers. Appl. Sci. **14**, 6901 (2024)

15. Hasholt, M.T., Jensen, O.M., Kovler, K., Zhutovsky, S.: Can superabsorbent polymers mitigate autogenous shrinkage of internally cured concrete without compromising the strength? Constr. Build. Mater. **31**, 226–230 (2012)
16. Sun, B., Wu, H., Song, W., Li, Z., Yu, J.: Hydration, microstructure and autogenous shrinkage behaviours of cement mortars by addition of superabsorbent polymers. Front. Struct. Civil Eng. 1–11 (2020)
17. Yang, J., Sun, Z., De Belie, N., Snoeck, D.: Mitigating plastic shrinkage cracking in alkali-activated slag systems by internal curing with superabsorbent polymers. Cem. Concr. Compos. **134**, 104784 (2022)
18. Mechtcherine, V., Secrieru, E., Schröfl, C.: Effect of superabsorbent polymers (SAPs) on rheological properties of fresh cement-based mortars—Development of yield stress and plastic viscosity over time. Cem. Concr. Res. **67**, 52–65 (2015)
19. Boshoff, W.P., et al.: The effect of superabsorbent polymers on the mitigation of plastic shrinkage cracking of conventional concrete, results of an inter-laboratory test by RILEM TC 260-RSC. Mater. Struct. **53**, 79 (2020)
20. Mechtcherine, V., et al.: Effect of superabsorbent polymers (SAP) on the freeze-thaw resistance of concrete: results of a RILEM interlaboratory test. Mater. Struct. **50**, 14 (2017)
21. Lee, H.X.D., Wong, H.S., Buenfeld, N.R.: Potential of superabsorbent polymer for self-sealing cracks in concrete. Adv. Appl. Ceram. **109**(5) (2010)
22. Snoeck, D., Van Tittelboom, K., Steuperaert, S., Dubruel, P., De Belie, N.: Self-healing cementitious materials by the combination of microfibres and superabsorbent polymers. J. Intell. Mater. Syst. Struct. **25**(1) (2014)
23. Snoeck, D., Schaubroeck, D., Dubruel, P., De Belie, N.: Effect of high amounts of superabsorbent polymers and additional water on the workability, microstructure and strength of mortars with a water-to-cement ratio of 0.50. Constr. Build. Mater. **72**, 148–157 (2014)
24. Mechtcherine, V., et al.: Effect of internal curing by using superabsorbent polymers (SAP) on autogenous shrinkage and other properties of a high-performance fine-grained concrete: results of a RILEM round-robin test, TC 225-SAP. Mater. Struct. **47**(3) (2014)
25. Snoeck, D., Pel, L., De Belie, N.: The water kinetics of superabsorbent polymers during cement hydration and internal curing visualized and studied by NMR. Sci. Rep. **7**, 9514 (2017)
26. Schröfl, C., Mechtcherine, V., Gorges, M.: Relation between the molecular structure and the efficiency of superabsorbent polymers (SAP) as concrete admixture to mitigate autogenous shrinkage. Cem. Concr. Res. **42**(6) (2012)
27. Mechtcherine, V., et al.: Testing superabsorbent polymer (SAP) sorption properties prior to implementation in concrete: results of a RILEM round-robin test. Mater. Struct. **51**, 28 (2018)
28. Danish, A., Mosaberpanah, M.A., Salim, M.U.: Robust evaluation of superabsorbent polymers as an internal curing agent in cementitious composites. J. Mater. Sci. **56**, 136–172 (2021)
29. Wyrzykowski, M., Lura, P.: Controlling the coefficient of thermal expansion of cementitious materials—A new application for superabsorbent polymers. Cem. Concr. Compos. **35**(1), 49–58 (2013)
30. De Schutter, G., et al.: Active control of properties of fresh and hardening concrete. RILEM Tech. Lett. **8**, 198–203 (2024)
31. Persson, B.: Experimental studies on shrinkage of high-performance concrete. Cem. Concr. Res. **28**(7), 1023–1036 (1998)
32. Yu, C.W., Bull, J.W.: Durability of Materials and Structures in Building and Civil Engineering. Whittles Publishing, Scotland (2006)
33. Li, K., Li, L.: Crack-altered durability properties and performance of structural concretes. Cem. Concr. Res. **124**, 105811 (2019)
34. Korda, E.: Active control of concrete curing by acoustic emission. Ph.D, Vrije Universiteit Brussel and Ghent University (2025)
35. Grosse, C.U., Ohtsu, M., Aggelis, D.G., Shiotani, T.: Acoustic Emission Testing: basics for Research—Applications in Engineering, 2nd edn. Springer, Cham (2021)
36. Hodulákova, M., Topolář, L.: The assessment of the structural changes of fine-grained cement-based composites in the early-age using acoustic testing. Int. Rev. Appl. Sci. Eng. **10**(1), 23–28 (2019)

37. Assi, L., Soltangharaei, V., Anay, R., Ziehl, P., Matta, P.: Unsupervised and supervised pattern recognition of acoustic emission signals during early hydration of Portland cement paste. Cem. Concr. Res. **103**, 216–225 (2018)
38. McLaskey, G.C., Glaser, S.D., Grosse, C.U.: Integrating broadband high-fidelity acoustic emission sensors and array processing to study drying shrinkage cracking in concrete. In: Sensors and Smart Structures Technologies for Civil, Mechanical, and Aerospace Systems, pp. 150–161, San Diego (2007)
39. Korda, E., De Schutter, G., Aggelis, D.G.: Acoustic signatures of hydration and microcracking in early-age concrete. Dev. Built Environ. **17**, 100353 (2024)
40. Lefever, G., et al.: Combined use of superabsorbent polymers and nanosilica for reduction of restrained shrinkage and strength compensation in cementitious mortars. Constr. Build. Mater. **251**, 118966 (2020)
41. Dzaye, E.D., De Schutter, G., Aggelis, D.G.: Study on mechanical acoustic emission sources in fresh concrete. Arch. Civil Mech. Eng. **18**, 742–754 (2018)
42. Lefever, G., Snoeck, D., De Belie, N., Van Vlierberghe, S., Van Hemelrijck, D., Aggelis, D.G.: The contribution of elastic wave NDT to the characterization of modern cementitious media. Sensors **20**(10) (2020)
43. Korda, E., Tsangouri, E., Snoeck, D., De Schutter, G., Aggelis, D.G.: Active control of concrete curing monitored by acoustic emission. In: Proceedings of the RILEM Spring Convention and Conference 2024. RSCC 2024, Conference on Advanced Construction Materials and Processes for a Carbon Neutral Society, Milano, pp. 412–420. Springer RILEM Publications S.A.R.L., Milano (2025)
44. Korda, E., Cousture, A., Tsangouri, E., Snoeck, D., De Schutter, G., Aggelis, D.G.: Active SAP desorption control in concrete through acoustic emission for optimized curing. Cem. Concr. Compos. **160**, 106067 (2025)
45. Iliopoulos, S.N., El Khattabi, Y., Aggelis, D.G.: Towards the establishment of a continuous nondestructive monitoring technique for fresh concrete. J. Nondestr. Eval. **35**(3) (2016)
46. Chung, K., Jeong, U., Oh, J.E.: Effects of magnetic field input cycle and peptizer on the MR effect of magneto-rheological elastomer based on natural rubber. Polym. Eng. Sci. **55**(11), 2669–2675 (2015)
47. Genç, S., Phulé, P.P.: Rheological properties of magnetorheological fluids. Smart Mater. Struct. **11**(1), 140–146 (2002)
48. Bossis, G., Volkova, O., Lacis, S., Meunier, A.: Magnetorheology: fluids, structures and rheology. In: Ferrofluids, pp. 202–230 (2002)
49. Rich, J.P., Doyle, P.S., McKinley, G.H.: Magnetorheology in an aging, yield stress matrix fluid. Rheol. Acta **51**(7), 579–593 (2012)
50. Arsenault, J.: Magnetorheological fluid dampened dowels for rigid pavement distress mitigation. Worcester Polytechnic Institute (2013)
51. Yang, G., Pan, J., Wang, D.: A review on the magnetorheological materials and applications. Int. J. Appl. Electromagnet Mech **75**(4), 407–443 (2024)
52. Ahamed, R., Choi, S.B., Ferdaus, M.M.: A state of art on magneto-rheological materials and their potential applications. J. Intell. Mater. Syst. Struct. **29**(10), 2051–2095 (2018)
53. Koo, J.H.: Using magneto-rheological dampers in semiactive tuned vibration absorbers to control structural vibrations. Virginia Polytechnic Institute and State University (2003)
54. Li, R., Zhou, H., Li, X., Zhang, Z., Wang, X.: Human simulated intelligent control on magneto-rheological vibration isolation system for subway floating slab track with moving load. In: 12th World Congress on Intelligent Control and Automation (WCICA), Guilin (2016)
55. Wei, K., Zhao, Z., Du, X., Li, H., Wang, P.: A theoretical study on the train-induced vibrations of a semi-active magneto-rheological steel-spring floating slab track. Constr. Build. Mater. **204**, 703–715 (2019)
56. da Silva Brandão, F., de Almeida, A.K., Miguel, L.F.F.: Methodology to minimize the dynamic response of tall buildings under wind load controlled through semi-active magneto-rheological dampers. Fracture Struct. Integr. **18**(69), 89–105 (2024)

57. Hou, Z., Li, Z., Wang, J.: Electrical conductivity of the carbon fiber conductive concrete. J. Wuhan Univ. Technol. **22**(2), 346–349 (2007)
58. Chen, B., Wu, K., Yao, W.: Conductivity of carbon fiber reinforced cement-based composites. Cem. Concr. Compos. **26**(4), 291–297 (2004)
59. Hayes, S.A., Lafferty, A.D., Altinkurt, G., Wilson, P.R., Collinson, M., Duchene, P.: Direct electrical cure of carbon fiber composites. Adv. Manuf.: Polym. Compos. Sci. **1**(2), 112–119 (2015)
60. Xie, N., Shi, X., Feng, D., Kuang, B., Li, H.: Percolation backbone structure analysis in electrically conductive carbon fiber reinforced cement composites. Compos. B Eng. **43**(8), 3270–3275 (2012)
61. Lu, D., Huo, Y., Jiang, Z., Zhong, J.: Carbon nanotube polymer nanocomposites coated aggregate enabled highly conductive concrete for structural health monitoring. Carbon **206**, 340–350 (2023)
62. Luo, J., et al.: Influences of multi-walled carbon nanotube (MCNT) fraction, moisture, stress/strain level on the electrical properties of MCNT cement-based composites. Sens. Actuators A **280**, 413–421 (2018)
63. Oumer, A., Lee, C., Ahn, E., Gwon, S.: Review on self-heating electrically conductive cementitious composites: focus on deicing and electrical curing. Constr. Build. Mater. **439**, 137232 (2024)
64. Li, H., Zhang, Q., Xiao, H.: Self-deicing road system with a CNFP high-efficiency thermal source and MWCNT/cement-based high-thermal conductive composites. Cold Reg. Sci. Technol. **86**, 22–35 (2013)
65. Lai, Y., Liu, Y., Ma, D.: Automatically melting snow on airport cement concrete pavement with carbon fiber grille. Cold Reg. Sci. Technol. **103**, 57–62 (2014)
66. Zhao, H.M., Wang, S.G., Wu, Z.M., Che, G.J.: Concrete slab installed with carbon fiber heating wire for bridge deck deicing. J. Transp. Eng. **136**(6), 500–509 (2010)
67. Kovtun, M., Ziolkowski, M., Shekhovtsova, J., Kearsley, E.: Direct electric curing of alkali-activated fly ash concretes: a tool for wider utilization of fly ashes. J. Clean. Prod. **133**, 220–227 (2016)
68. Tuan, C.Y., Yehia, S.A.: Implementation of conductive concrete overlay for bridge deck deicing at Roca, Nebraska. In: Civil Engineering Faculty Proceedings & Presentations, p. 17 (2004)
69. Goudjil, N., Djelal, C., Vanhove, Y., Kada, H., Heloun, N.: Impact of temperature on the demoulding of concrete elements with a polarization process. Constr. Build. Mater. **54**, 402–412 (2014)
70. Goudjil, N., Vanhove, Y., Djelal, C., Kada, H.: Electro-osmosis applied for formwork removal of concrete. J. Adv. Concr. Technol. **110**(9), 301–312 (2012)
71. Choi, M.S., Kim, Y.S., Kim, J.H., Kim, J.S., Kwon, S.H.: Effects of an externally imposed electromagnetic field on the formation of a lubrication layer in concrete pumping. Constr. Build. Mater. **61**, 18–23 (2014)
72. Wang, X., Li, Z., Han, B., Yu, X., Zeng, S., Ou, J.: Intelligent concrete with self-x capabilities for smart cities. J. Smart Cities **2**(2) (2016)
73. Wang, Q., Snoeck, D.: To boldly go where no one has gone before: sulfur concrete, a promising construction material fulfilling the demands for a sustainable future on celestial objects: a review. Mater. Today **72**, 301–317 (2024)
74. Ayar, P., Moreno-Navarro, F., Rubio-Gámez, M.C.: The healing capability of asphalt pavements: a state of the art review. J. Clean. Prod. **113**, 28–40 (2016)
75. Xu, S., García, A., Su, J.F., Liu, Q., Tabakovic, A., Schlangen, E.: Self-healing asphalt review: from idea to practice. Adv. Mater. Interfaces **5**, 1800536 (2018)
76. Wan, P., et al.: Dual responsive self-healing system based on calcium alginate/Fe_3O_4 capsules for asphalt mixtures. Constr. Build. Mater. **360**, 129585 (2022)
77. Sun, D., et al.: A comprehensive review on self-healing of asphalt materials: mechanism, model, characterization and enhancement. Adv. Coll. Interface. Sci. **256**, 65–93 (2018)
78. Li, C., Xiao, M., Dong, J., Ren, J., Guo, X.: Study on the factors affecting the self-healing performance of asphalt mixture and pavement based on fracture mechanics and calculation formula. Theoret. Appl. Fract. Mech. **126**, 103954 (2023)

79. Liang, B., Lan, F., Shi, K., Qian, G., Liu, Z., Zheng, J.: Review on the self-healing of asphalt materials: mechanism, affecting factors, assessments and improvements. Constr. Build. Mater. **226**(A), 120453 (2021)
80. Patti, F., Mansour, K., Pannirselvam, M., Giustozzi, F.: Mining materials to generate magnetically-triggered induction healing of bitumen on smart road pavements. Constr. Build. Mater. **171**, 577–587 (2018)
81. Jeoffroy, E., Bouville, F., Bueno, M., Studart, A.R., Partl, M.N.: Iron-based particles for the magnetically-triggered crack healing of bituminous materials. Constr. Build. Mater. **164**, 775–782 (2018)
82. Jeoffroy, E., Koulialias, D., Yoon, S., Partl, M.N., Studart, A.R.: Iron oxide nanoparticles for magnetically-triggered healing of bituminous materials. Constr. Build. Mater. **112**, 497–505 (2016)
83. Zhao, H., Zhong, S., Zhu, X., Chen, H.: High-efficiency heating characteristics of ferrite-filled asphalt-based composites under microwave irradiation. J. Mater. Civil Eng. **29**(6) (2017)
84. Zhu, X., Cai, Y., Zhong, S., Zhu, J., Zhao, H.: Self-healing efficiency of ferrite-filled asphalt mixture after microwave irradiation. Constr. Build. Mater. **141**, 12–22 (2017)
85. Franesqui, M.A., Yepes, J., García-González, C.: Top-down cracking self-healing of asphalt pavements with steel filler from industrial waste applying microwaves. Constr. Build. Mater. **149**, 612–620 (2017)
86. Liu, Q., Li, B., Schlangen, E., Sun, Y., Wu, S.: Research on the mechanical, thermal, induction heating and healing properties of steel slag/steel fibers composite asphalt mixture. Appl. Sci. **7**(10) (2017)
87. Malekmohammadi, A., Farzadnia, N., Hajrasouliha, A., Mayer, A.L.: Sensing systems in construction and the built environment: review, prospective, and challenges. Sensors **23**(24), 9632 (2023)
88. Han, B., Ding, S., Yu, X.: Intrinsic self-sensing concrete and structures: a review. Measurement **59**, 110–128 (2015)
89. Tsangouri, E., Karaiskos, G., Deraemaeker, A., Van Hemelrijck, D., Aggelis, D.: Assessment of acoustic emission localization accuracy on damaged and healed concrete. Constr. Build. Mater. **129**, 163–171 (2016)
90. Lefever, G., Charkieh, A.S., Abbass, M., Van Hemelrijck, D., Snoeck, D., Aggelis, D.G.: Ultrasonic evaluation of self-healing cementitious materials with superabsorbent polymers: mortar versus concrete. Dev. Built Environ. **13**, 100112 (2023)
91. Xu, N., et al.: Employing ultrasonic wave as a novel trigger of microcapsule self-healing cementitious materials. Cem. Concr. Compos. **118**, 103951 (2021)

Summary and Further Perspectives Regarding Active Control of Cementitious Materials

Geert De Schutter and Jay Sanjayan

Abstract This chapter is the concluding chapter of the RILEM State-of-the-Art Report on Active Control of Properties of Fresh and Hardening Cementitious Materials. Active control of properties of cementitious materials refers to the action or intervention for controlling some specific properties (e.g. rheology, stiffening) by applying an external trigger signal that induces some response in the cementitious material, with a modification of properties or behaviour as a result. This final chapter summarizes the information given in the full report, and presents some further perspectives, including remaining challenges.

Keywords Active control · Cementitious materials · Fresh properties · Hardening properties

1 Introduction

During the execution phase of concrete structures, the construction processes are prone to variations amongst others linked to environmental conditions and fluctuations in properties of constituent materials. In current approaches within industry, insufficient methodologies are available to deal with such variations, too often leading to practical challenges and/or reduced final quality. As an example, due to variations in environmental conditions, the workability of fresh concrete might be influenced in an unfavourable way, potentially leading to difficulties with formwork filling and compaction of the concrete.

G. De Schutter (✉)
Department of Structural Engineering and Building Materials, Magnel-Vandepitte Laboratory, Ghent University, Ghent, Belgium
e-mail: Geert.DeSchutter@ugent.be

J. Sanjayan
Swinburne University of Technology, Melbourne, Australia

G. De Schutter and J. Sanjayan (eds.), *Active Control of Properties of Fresh and Hardening Cementitious Materials*, RILEM State-of-the-Art Reports 50,
https://doi.org/10.1007/978-3-032-30340-0_10

Long-term practical experience has provided some answers for some challenges. Casting and compacting for instance can be practically controlled by poker vibration, while curing can be improved by providing heat and/or moisture. This kind of practical methodologies help to overcome natural variations in materials and in construction conditions. Such interventions (vibration and curing) are in fact active controls. Recently, many new methods of active control have been researched, and useful and practical outcomes have been demonstrated. This state-of-the-art report discusses all the different technologies that can be classified as active controls—a relatively new terminology. By bringing them together in one umbrella term of "active controls" and providing comprehensive discussions of the technologies with rational classification of the technologies, this report provides a scientific framework for further developments and industry implementations of these technologies.

In order to improve the robustness of construction processes in concrete industry and on site, active control is now comprehensively researched. **Active control of properties of cementitious materials refers to the action or intervention for controlling some specific properties (e.g. rheology, stiffening, …) by applying an external trigger signal that induces some response in the cementitious material, with a modification of properties or behaviour as a result.** In the ideal active control process, the intervention is based on observations and/or measurements (quantified) giving the necessary information to take a decision regarding required intervention. A feedback loop between measurement and signal seems an optimal situation.

Some historical methodologies that could be considered as active control 'avant la lettre' have been summarized in Chap. "Introduction to Active Control of Properties of Fresh and Hardening Cementitious Materials". They can be based on different types of intervention: mechanical, chemical, thermal, hygral, pressure, magnetic, electric, microwave. In more recent approaches, the active control mechanism is providing extra opportunities by the addition of responsible polymers or particles to the concrete mix design, that can be triggered by an external signal. As an example, electro- or magneto-responsive superplasticizers can be considered to modify the flowability of the concrete during casting in an active way, on demand. The new approaches give ample opportunity to control properties and/or to control processes.

Control of properties can in first instance be related to rheology, setting, and shrinkage. The literature regarding active control of those properties has been synthesized and summarized in Chaps. "Active Control of Rheology of Cementitious Materials" (rheology), "Active Control of Setting of Cementitious Materials" (setting) and "Active Control of Concrete Shrinkage and Associated Cracking" (shrinkage). A short summary is given hereafter in Sect. 2.

Some processes offer very good opportunities for the application of active control methodologies. For some very relevant processes, available literature information has been summarized in Chaps. "Active Control of Concrete Pumping" (pumping), "Active Control of Formwork Casting of Cementitious Materials" (formwork casting), "Active Control of 3D Printing of Cementitious Materials" (3D concrete printing) and "Active Control of Cementitious Slabs and Roads" (slabs and roads). A short summary is given hereafter in Sect. 3.

2 Active Control of Properties

The fresh concrete properties, such as rheology and setting, are highly suited for active control since the properties can be measured, and based on the feedback, interventions can be made to adjust the properties. Early age and hardened concrete properties, such as shrinkage and strength, can also be controlled by interventions such as active curing but are less amenable for feedback control.

2.1 Rheology

Chap. "Active Control of Rheology of Cementitious Materials" of this state-of-the-art report has systematically reviewed and compiled the available literature information dealing with the active control of rheological properties of fresh cementitious materials. First, a short introduction has been given to some fundamental rheological definitions and concepts, in order to guide the reader in further understanding the active rheology control concepts.

The subfield of active rheology control (also called ARC) of fresh cementitious materials has substantially evolved during the last decade. Several control signals can be considered, with published information regarding control by magnetic fields, electric fields, vibration, microwaves, temperature, ultrasound, pH and pressure. For each control methodology, some background information has been given, including underlying principles, influencing parameters, and potential applications to construction practice.

Some recent developments rely on the use of switchable superplasticizers, e.g. electro- or magnetoresponsive polycarboxylate ethers containing an additional type of side chains with specific functionalities. Other options rely on responsive nano- or microparticles, e.g. magnetizable SCMs or fillers or even fine sand particles, turning the fresh cementitious material into a magnetorheological fluid. Although several concepts have been successfully demonstrated at paste or mortar level in laboratory conditions, further valorisation and upscaling to realistic conditions within construction industry is still going on.

2.2 Setting

Although magnetorheological approaches as described in Chap. "Active Control of Rheology of Cementitious Materials" are also applicable to obtain setting-on-demand, also referred to as active stiffening control (ASC), chemically induced setting has been more intensely studied and reported in literature up to now. It can be particularly relevant for processes like 3D printing (see also further). Most papers report the application of inline introduction of a chemical accelerator, combined with an inline mixer, mostly a dynamic mixer although a static mixer can also be considered. The main focus in literature is on the mixture design, going hand in hand with the choice of an appropriate accelerator.

The choice or design of the inline mixer received less attention in literature. Even with low mixing efficiency, literature reports that fast setting rates can be achieved in practice. However, it has to be considered that inadequate mixing can induce quality loss of the long-term properties of the (hardened) concrete. This leaves ample opening for further optimization studies of inline mixers, either dynamic or static.

Next to chemical setting-on-demand, heat-induced setting control has also been extensively reported, e.g. by means of microwaves or electric heating. Although uniform heating can be obtained with shorter heating durations, the adoption of microwave technology in the construction industry is not straightforward because of safety concerns. A potential breakthrough can maybe come soon integrating microwave technology in print heads for 3D concrete printing (see also Sect. 3.3).

A combined application of two or more signals can be very efficient in view of setting-on-demand. As an example, a combination of a chemical accelerator and a heat signal at a specific moment can result in very fast controlled setting.

2.3 Shrinkage

Shrinkage and associated cracking are challenges in concrete construction practice. The main mechanisms, including chemical shrinkage, self-desiccation, carbonation shrinkage, plastic shrinkage and drying shrinkage are briefly recapitulated in Chap. "Active Control of Concrete Shrinkage and Associated Cracking". The importance of curing conditions, and potential mitigation measures such as the application of superabsorbent polymers (SAP) are briefly summarized as well.

Based on the explained shrinkage mechanisms, Chap. "Active Control of Concrete Shrinkage and Associated Cracking" included a brief and preliminary overview of the properties that could be actively controlled in view of reducing shrinkage and the associated cracking risk. For a more detailed discussion of the active control mechanisms related to shrinkage, reference is made to Chap. "Active Control of Cementitious Slabs and Roads", focussing on the case of concrete slabs and roads.

3 Active Control of Applications

Active control can potentially be applied to many situations in construction where an intervention may be needed to achieve certain outcomes. Four main applications have been highlighted in this report, where active control can be particularly beneficial, namely, pumping, mould/formwork casting, 3D printing and slabs and roads. Other applications where active controls are currently being used or have the potential to be used include shotcreting, slip/jump forming (also recently smart dynamic casting), under water concreting etc., that can be explored further.

3.1 Pumping

On a daily basis, large volumes of fresh concrete are pumped on construction sites or in precast industry. Due to the high pressures involved and the heterogeneity of concrete and susceptibility of segregation, pumping practice is often a challenging and high risk operation. Small variations in material properties can lead to insufficient pumpability, or even to complete blocking of pumping pipes.

For a given pressure loss per unit of pipe length, the linear evolution of shear stress within the pipe does not depend on the rheological properties of the fresh concrete. However, velocity profile and resulting discharge rate significantly depend on the material, as well as on phenomena like shear-induced particle migration leading to the formation of a lubrication layer. By actively modifying the rheological properties of the fresh concrete (see also Chap. "Active Control of Rheology of Cementitious Materials"), the pumping pressure-discharge relation can in principle be modified as well. This pumping control could be based e.g. on magneto-rheological principles. First ideas are available in literature on how to actively control pumping of fresh cementitious materials, as mentioned in Chap. "Active Control of Concrete Pumping". However, further research is ongoing in order to bring the pumping control technology closer to practical applications in real conditions.

3.2 Formwork/Mould Casting

Formwork or mould casting still is the most popular way to produce concrete elements and structures, with a variety of formwork types that can be considered to support the fresh concrete until it is sufficiently hardened to be self-supporting. In case of formwork casting, practical issues and problems can occur related to improper filling, excess formwork pressure, formwork leakage, and damage while demoulding.

Chapter "Active Control of Formwork Casting of Cementitious Materials" first refreshed some basic information regarding formworks. Afterwards, the chapter has given an overview of literature information regarding active control options that can be beneficial in case of formwork casting. This includes active rheology and stiffening control to facilitate filling and reduce formwork pressure, magneto-rheological leaching control, and electric demoulding control.

3.3 3D Printing

3D concrete printing is a remarkable process that could benefit from active control strategies in many ways. Typically, the printing process combines different subsequent steps, including pumping, extruding and building of the layers. Active control actions can help in achieving appropriate levels of pumpability, extrudability and buildability of the same material at different phases of the printing process, as was summarized in Chap. "Active Control of 3D Printing of Cementitious Materials".

During the pumping stage, active rheology control through vibration and magnetic intervention can be adopted to ensure a stable flow without blocking or segregation. After pumping, hydration and setting control is relevant while the material is within the print head or nozzle. At this stage, chemical control and/or temperature control can be helpful, making sure that the material reaches a very high buildability while building the layers of the printed element. Magneto-rheological approaches are also very relevant here, with even a higher buildability in case of so-called hard magnetic particles with remanent magnetization properties. Whatever control technique is used to increase the buildability, due attention should also be given to the interlayer adhesion.

Active control of plastic and drying shrinkage of the printed element is a further step in increasing the quality of the final structure, avoiding cracking issues as observed in some printed elements. In general, applying various active control methodologies at different stages of the 3D printing process, contributes to the final quality and service life of the printed structure.

3.4 Slabs and Roads

Concrete slabs and roads are typically prone to early age cracking. Based on capillary pressure monitoring, active control of plastic shrinkage cracking can be achieved by well-timed and temporarily rewetting, making curing more economical compared with continuously fog spraying. Further work, however, should be done in view of obtaining a standardized method for the determination of the capillary pressure threshold.

As currently studied, active control of internal curing can be achieved by activating external curing as soon as SAP particles start releasing their stored water. This moment can be traced by real-time monitoring using acoustic emission. This approach can lead to reduced SAP dosages, while reducing early-age shrinkage and microcracking. This contributes to increased sustainability and reduced cost of concrete roads and slabs.

Magnetorheological materials provide additional opportunities for responsive civil engineering solutions, enabling dynamic control over vibration, stability and traction, as demonstrated in simulation and laboratory testing. Real-world applications, however, require further valorisation steps.

In future, the integration of active control mechanisms, automation, and artificial intelligence opens the door for further innovations, including structural health monitoring and self-maintenance, contributing to sustainable infrastructure with reduced resource consumption and environmental impact.

4 Benefits of Active Control

In construction industry in general, and concrete industry more specifically, the quality and performance of the final structure or of the structural element typically depend on practical and environmental conditions during the execution. Weather conditions typically influence the working conditions on site, impacting the activities of the workers and impacting the properties of the materials as well, especially of fresh cementitious materials with temperature and humidity dependent performance. In current practice, only limited options are available to adjust non-optimal material performance. Typically, the material passively shows its properties like workability and setting as influenced by the environmental conditions. Active control provides a unique way of effective and efficient interventions to bring the material back to more optimal performance in case of deviations e.g. due to environmental conditions or natural variations in production or processing. In this way, active control of properties of fresh and hardened cementitious materials can bring manifold benefits:

- Better control of material properties
- Better control of processing steps
- Better control of flow and hardening of the cementitious material
- Better crack control during hardening and drying
- Reduced level of construction errors or unwanted deviations
- Lower risk of rejection
- Overall better quality of the final structure.

These benefits can hopefully compensate for the higher cost that might be linked to the active control technology, requiring responsive materials in combination with the trigger signals. Overall cost studies, however, are not yet available, as the active control options have not been deployed yet in current construction industry.

5 Remaining Challenges for Active Control

Active control methodologies for fresh and hardening cementitious materials are currently still having a low technology readiness level (TRL). Most results and proofs of concept have currently been obtained in laboratory conditions, without real upscaling yet to larger scale in industrial environments. This upscaling is an important remaining challenge, including several aspects to be further studied:

- Upscaled production of responsive materials, e.g. responsive polymers, at an acceptable production cost
- Upscaled production of concrete with responsive materials, without destroying responsive materials during mixing stage
- How to practically organize the required trigger signals at the larger scale of the construction site or precast factory?
- How to trigger larger volumes of concrete, while the penetration of the signal might be limited?
- What about safety of workers in case of electric or magnetic signals given at larger scale?
- What about acceptance of the innovative controllable materials within the current framework of traditional standards and prescriptions?
- How to train workers to deal with innovative materials in a good way?
- Which parties are performing which tasks during processing? Who is deciding on required active control actions?
- What about responsibilities in case of actively controlled materials?

Currently, many valorisation and upscaling studies are ongoing. In many cases, this is still at academic research level, in protected research environments. Gradually, industry needs to be more and more involved in the valorisation studies, validating the active control methodologies in real environment and at real scale. After successful validation in industry, cost assessments and business plans will be needed. The remaining challenges are significant, indeed. However, combined efforts can bring construction industry also to levels of increased efficiency, quality and safety.

6 Conclusion

This RILEM State-of-the-art Report on "Active Control of Fresh and Hardening Cementitious Materials" introduced a new field within concrete research. By means of responsive particles, minerals and/or polymers, the behaviour of the fresh and hardened cementitious material can be triggered to a required level of adjustment. Trigger signals can be diverse, e.g. electric, magnetic, chemical. The (historical) background of active control methodologies have been illustrated, and inspiring approaches from other fields have been mentioned as a source of inspiration.

Active control of cementitious materials can be approached with focus on the properties (e.g. rheology, setting shrinkage) or with focus on processes (e.g. pumping, formwork casting, 3D printing, production of slabs and roads). This report describes the innovative active control methodologies from both points of view. Theoretical background is summarized, and a comprehensive literature overview is given, with 2025 as the reference year.

It is clear that active control can have significant benefits for construction industry. With active tools and methodologies to better master the behaviour of the fresh and hardening concrete as well as the different processing steps, improved final quality can be achieved with less failures and reduced number of rejections. This will contribute to the overall quality and sustainability within the construction industry.

However, it is also clear that the active control methodologies are still in relatively premature stage, requiring significant further studies, including upscaling studies to bring the methods from laboratory to industry. Several novel and innovative technologies have been explored for active control by researchers in laboratory conditions. Some of these technologies have the potential to solve many of the current problems faced by industry.

Kolibri 360 GmbH
Ferdinand-Jühlke-Straße 7
99095 Erfurt, Deutschland
produktsicherheit@kolibri360.de